TECHNOLOGY BRINGS POSITIVE OR NEGATIVE INFLUENCE

TO HUMAN BEHAVIOR

JOHN LOK

Contents

Preface

Introduction

Nowadays, robotic had been applies to different aspect nu humans, they may include any hospitals' surgeon rooms medical surgeon equipment aspect, hotel room food delivery or hotel front line customer service aspect, shopping center, leisure places cleaning task aspect, student educational service aspect, warehouse logistic goods transport delivery tasks aspect , even general restaurants kitchen cooking tasks aspect, lawyer and accountant firms general bookkeeping, law writing draft clerical tasks aspects etc. Can robotic invention help employers to improve efficiencies or raise productivities in order to raise economic growth or it can be replaced low-skillful workers to cause any low-skillful jobs are done by artificial robots to achieve many low skillful workers will lose jobs and unemployment ratio will be influenced to raise in our societies?

How social change influences human behavioral change ? Why human behavior may be influenced by social change? Our individual behavior whether can be influenced to bring negative or positive attitude by social change? I shall attempt to indicate cases to explain whether our individual behavior can be influenced to changed by social environment change. Readers can have more understand how and why social change may influence our behavior in possible. Behavioral economy is one useful and fun social subject. Behavioral economists ususally research how and why human behaviors may influence economy growth or recession, or how and why economy environment changing factor may influence human behavior changes.

In my this book, I shall attempt to explain how and why ecommerce may be one kind network human job. Also, I shall indicate reasons to explain why human network behavior may bring direct or indirect influences to economy growth or recession in our global societies in macro and micro economy view. Why leisure changing environment may influence human behavior , even economic environment changes. I shall indicate cases to explain any possible human social activities may bring direct or indirect influences to cause our social economic growth or recession in consequency in possible. I hope that my readers can feel more understanding whether what real meaning of behavioral economy is the relationship between our behaviors and our economy.

I write this book aims to discuss whether future robotic invention may help our societies improve economic growth or it can bring recession. This book includes two parts to discuss, this part indicates what reasons to explain why robotics may influence our societies economic growth as well as second part indicates what reasons to explain why robotics may bring our social recession both in possible. Readers can make individual analysis to judge whether robotics can influence our social economic growth or it may bring recession more.

Prologue

Contents

CHAPTER I

(AI) raises driven automation industry development

1.1 (AI) - driven automation industry development how to influence work nature change

(AI) -driven automation industry will create wealth and expand economy growth to any countries, but it will be accompanied by changed in the skills that workers need to learn. One of main ways that technology increases productivity is by decreasing the number of labor hours needed to create a unit of output. It implies (AI) technology will influence low educated and low skillful labor number to be decreased (reduction employment number).

In contrast, technological change tended to work in a different direction throughout the nowadays. The advance of computer and the internet raised the relative productivity of higher skilled workers. So, routine-intensive occupations that focused on predictable tasks disappearance, such as switch board, operators, filming checkers, travel agents and assembling line workers etc. were particularly replaced by new technologies.

However, today, it may be challenging to predict exactly which jobs will be most immediately affected by (AI) driven-automation. The reason is because (AI) is not a single technology, but rather a collection of technologies that are felt unevenly through the economy to influence job changing both negatively and positively. In positively view point, (AI) driven-automation will make many workers more productive and increase demand for certain skills. Consequently, new jobs are likely to be directly create in areas , such as the development and supervision of (AI) as well as indirectly created in a range of areas throughout the economy as higher incomes lead to expanded demand. Otherwise, in negatively view point, many traditional human needed (demand) skillful jobs will be threatened by automation are highly concentrated among lower-paid, lower-skilled and less -educated workers. It means automation will cause pressure on demand for this group, pressure and employment, if (AI) can replace the low skilled and less educated workers' jobs. Thus, (AI) will have negative influence to impact on the labor market.

(AI) capabilities will enable automation of some tasks that have long required human labor. Why can (AI) replace some simple human jobs? For example, advances in robotics are expanding machines' abilities to interact with and sharp the physical world. Combined , (AI) and robotics will give rise to smarter machines that can perform more sophisticated functions than ever before and brings more advantages that humans have exercised. This will permit automation of many tasks now performed by human workers and could change the shape of the labor market and human activity.

1.2 How (AI) influences labor market

Today, it may be challenging to predict exactly which jobs will be most immediately affected by (AI)-driven automation. Because (AI) is not a single technology, but rather a collection of technologies that are applied to specific tasks.

Some specific predictions are possible based on the current (AI) technology. For example, driving jobs and house cleaning jobs, bank counter service jobs, telephone enquiry service operators. Restaurant cooking jobs, simple accounting record service jobs etc. that require relatively less education to perform. Advancements in computer vision and related technologies have made the feasibility of fully appear more likely, potentially displacing some workers in driving-dominant professions. Seemingly similar robot, for which the operational tasks is less specific of navigating to a specific destination when following a set of given rules and preserving safety.

In the future, the effects of (AI) on the labor market in the decade ahead will continue the trend toward skill-biased change that computerization and communication innovations have driven in recent decades. Thus, some human driving occupation will be disappeared or replaced by (AI) automation driven. For example, bus drivers, light truck or delivery services drivers, heavy and tractor-trailer truck drivers, school drivers, tax drivers, travel bus drivers.

However, (AI) technology could enable some workers to focus time on other job responsibilities, boosting their productivity, and actually raised wage growth among those still holding the reshaped jobs. For example, salespeople, who currently spend a considerable amount of time driving could find themselves able to do other work when a car drives them from place to place, or inspectors and appraisers could fill out paperwork, when their car drives itself. This (AI) -driven technology should make these workers more productive, with (AI) -driven technology serving as a complement, not a substitute. New jobs will also likely be created, both in existing occupations cheaper transportation costs with lower prices and increase demand for products and all the related occupations, such as service and fulfillment, and in new occupations not currently foreseeable. What kind of jobs will be created by (AI) technology? Predicting future job growth is extremely difficult, due to it depends on technologies or substitute for existing today as well as they may complement or substitute for existing human skills and jobs. However, (AI) will also lead to substantial indirect job creation to the degree it raises productivity and wages, it may also lead to higher consumption that would support additional jobs from high-end draft production to restaurant and retail. The future(AI) " augmented intelligence", the technology's role is as assisting and expanding the productivity of individuals rather than replacing human work. Thus, based on the biased-technical change framework, demand for labor will likely increase the most in the areas where humans complement (AI) automation technologies. For example, (AI) technology , such as IBM's Watson may improve early detection of some cancers or other illnesses, but a human healthcare professional is needed to work with patients to understand and translate patients' symptoms, inform patients of treatment options, and guide patients through treatment plans. Shipping companies may also partner workers who pick up and deliver products over the last feet with (AI) enabled autonomous vehicles that move workers efficiently from site to site. In such cases, (AI) augments what a human is able to do and allows individuals to either be move effective in their specially task or to operate on a larger scale. Thus, it seems (AI) technology will also create new jobs, raise productivities and workers' efficiencies.

Redefining management in the workforce of artificial intelligence

2.1 Change management

In the future, due to artificial intelligence influences to some kind of human jobs nature. So, the kind of human jobs of management methods will also need to change to adapt the artificial intelligence technology input to their organizations. It will cause challenges for every executive and manager if who won't have effort to manage their teams how to apply artificial intelligence technology to work efficiently and easily. For example, division of labor will change among humans and machines will increase. Thus, companies will have to adapt their training performance and talent strategies how to emphasize on work that how to make human judgment and skills and experimentation. Thus, (IA)'s greatest impact will be on administrative coordination and control tasks, such as scheduling , resource allocation.

In fact, mangers will encounter this challenges: How to apply human experience and expertise to judge critical business decisions and practices when the information available is insufficient to suggest a successful course of action? Due to this kind of work will require new skills and mindsets. I shall indicate these change management methods to adapt (AI) technology. Such as: administration and routine tasks, scheduling , allocation of resources and reporting will fall within the intelligence machines, responsibilities that have long been reserved for humans. For example, a typical store manager or a lead nurse at a nursing home most constantly arrange shift schedules, accounting for staff members' absences owing to illness, vacation time or sudden departures.

Thus, the managers need to learn how to arrange new division of labor within the organizations after (AI) technology had been implemented to the organization. Artificial intelligence is currently influencing into once considered exclusive to humans: assessing and acting on human emotions and personality traits. The influences to managers need to change their strategies to adapt (AI) technology implements include such as below:

Firstly, managers need to spend the bulk of their time on coordination and control tasks from intelligent system implements. Their time spending on these major three aspects from impact of intelligent system: coordinate and control, solve problems and collaborate and people and community , strategy and innovation three aspects. Thus (AI) will influence managers need to change their judgment method to teach whose teams how to adapt the (AI) system operations in any organizations.

Secondly, (AI) will influence top, middle and low level management needs to change to adapt the (AI) technology operations to any owned (AI) technology organizations in the future. Intelligent machines must be trained in context. Just like humans , on-the-job training is a requirement for such machines because they typically arrive with only very general capabilities. To get the most from (AI), managers at all levels must participate in the instructional experience and in the learning process and provides managers' familiarity with such systems on these aspects, e.g. How the system works and generate advice, how the system has a proven track record , how the system provides convincing explanations , how the system can make simple rule- based decisions.

Thirdly, managers need to learn how to make judgment more accurate (AI) systems assistance. Although (AI) will invariably take on more routine work and even augment human decision-making, it won't judgment work, the application of human experience and expertise to critical business decisions when the information available is insufficient to suggest a successful course of action or reliable enough to suggest an obvious course of action. For a sense of the nature of judgment work, consider big data marketing and sales analytics. Such analytics often provide insights that can inform promotional campaigns, including predicting which promotions will generate desired sales brand further into the future, marketing executives need use judgment, combining analytics with their own and others' insight and experience.

The application of experience and expertise to critical business decisions and practice represents the real value of human judgment. But, when artificial intelligent machines are implemented to any organizations to assist the low, middle and top level management to make any business judgment. These forms of judgment work that managers can gather data interpretation, idea development more absolute from (AI) machine assistance. Thus, why these level management executives need to learn how to apply (AI) machines to help them to make any business judgment more accurate.

2.2 How (AI) influences organizational change

Consequently creative and social intelligence will be in even greater demand as (AI) makes in management and the workforce. This development will represent a long term trend in labor markets , one characterized by intensifying demand and reward for social skills with a growing desire for creative capabilities, managers will seek to fashion of ideas and hypotheses from inside and outside of the enterprise to shape solutions to their most pressing business problems. Thus, (AI) will influence overall organizational team members who have chance to participate any decision to make more accurate business judgment.

Many managers mistakenly view judgment work as only an individual discipline, failing to appreciate that it can also involve decide interpersonal and organizational practices. In more complex settings, judgment is typically a collective outcome of individuals' and teams' diverse perspectives, insights and experiences. And often , the resulting choices are better informed than decisions that an individual would have arrived at on his or her own.

Thus, when any organizations apply (AI) technology to assist managers to gather data and ideas to make any judgment. In these cases, organizations can create the conditions for effective collective judgment by establishing structures , such as " shadow advisory boards" that prompt managers and employees to source and synthesize multiple perspectives. Thus, a traditional organization (firm) might freshen its thinking is t put together a shadow advisory board, comprised of young, digital people who can apply (AI) machine assistance to make judgment work more accurate whether related to people development, problem-solving or strategizing and innovating for considerable degrees of creative and social intelligence.

Thus, on the one hand, (AI) technology machine augmentation and automation can give these advantages to human (organization managers) , e.g. developing people and community, solving problems and collaborating, coordinating and controlling work, shaping strategy and leading innovation. Besides, on the other hand, the next generation managers need have these individual attitude to treat intelligent machines to be as colleagues.

When, judgment is a human skill, intelligent machines can accelerate human learning that supports it, assisting in data -driven simulations, scenarios and search and discovery activities. Focuses on judgment work, some decisions require insight beyond what data can tell them. This is the sweet sport for human judgment, the application of experience and expertise to critical business decisions and practices. Thus, managers will also need to find ways to learn how to use digital (AI) technologies to tap into the knowledge and judgment of partners, customer external stakeholders and role models in other industries after the (AI) machine had been implemented to the organization.

Future works change:Automation, employment and productivity

3.1 How (AI) influences employment

Human future " micro to macro" industry trends will be affected business strategy and public policy by (AI) technology. In the future (AI) technology will influence those six themes: productivity and growth, natural resources, labor markets, the evolution of global financial markets, the economic impact of technology and innovation and urbanization. However, (AI) technology will bring economic benefits of tackling gender inequality, a new global competition, Chinese innovation and digital globalization.

Nowadays, advances in robotics artificial intelligence, and machine learning are in a new age of automation, as machines match or outperform human performance in a development to any countries. For example, automation of activities can enable businesses to improve performance by reducing errors and improving quality and speed, and in some cases achieving outcomes that go beyond human capabilities. For example, some research indicated automation could raise productivity growth globally by 0.8 to 1.4 % annually; more than 2,000 work activities across 800 occupations. When less than 5% of all occupations can be automated using demonstrated technologies about 60% of all occupations have at least 30% of constituent activities that could be automated. Many occupations will change that will be automated away: Activities most susceptible to automation involve physical activities, in highly structured and predictable environments, as well as the collection and processing of data. They are most prevalent in manufacturing , accommodation and food service and retail trade and include some middle-skill jobs. For example, such as natural language processing is a key factor. Beyond technical feasibility, the cost of technology competition with labor including skills and supply and demand dynamics, performance benefits including and beyond labor cost savings, and social and regulatory acceptance will be affected by (AI) automation technology. Thus, (AI) automation will impact to influence global employment in those aspects as below:

Firstly, assuming that people are displaced by automation will find other employment. The anticipated shift in the activities in the labor force is of a similar order as the long-term shift away from agriculture and decreases in manufacturing share of employment. Both of manufacturing and agriculture industries which would be accompanied by the creation of new types of work not foreseen at the time.

Secondly, for business, the performance benefits of automation are relatively clear. Thus, the businessmen have opportunities for their micro economies to benefits from the productivity growth potential and macro economies to benefit to encourage continued progress and innovation , investment and market incentives. At the same time, employers must innovate policies to help workers and institutions adapt to the impact on employment.

This will likely include rethinking education and training, income support and safety nets , as well as support for those dislocated, when employees need to leave themselves homes to move to other cities to learn new (AI) automation works. Thus, individuals in the workplace will need to engage move comprehensively with machines as part of their everyday activities, and acquire new skills that will be in demand in the new automation age. Consequently , the scale of shifts in the labor force over many decades that automation technologies can be a similar order to the long -term technology -enables shifts in the developed countries' workforces away from agriculture in the 21 th century. Those shifts did not result in long-term mass unemployment because they were accompanied by the creation of new types of work not foreseen at the time. However, human will still be needed in the workforce when the total productivity gains are caused by (AI) technology.

3.2 What occupations will be influenced by (AI) technology.

In the future, scientists predict that these occupations will be influenced by (AI) technology mostly. They include : retail salespeople, food and beverage service workers, language or translation teachers, health practitioners. Since these work activities have a more relevant occupations are made up of a range of activities with different potential for (AI) automation . For example, a retail salesperson will spend more time interacting with customers, stocking shelves , or ringing up sales. Each of these activities is distinct and requires different capabilities to perform successfully.

Thus, these job activities have similar simple control characteristics. Simple activities include greet customers, answer questions about products and services, clean and maintain work areas, demonstrate product feature process sales and transactions. All these activities can have similar simple activities in order to (AI) machines can be learn how to do these activities from (AI) technology . For example, the capability perception includes sensory perception, cognitive capabilities, such as retrieving automation, recognizing known patterns(supervised learning), logical reasoning problem solving.

Thus, (AI) machine is such human, which has feeling and emotion, such as social and emotional sensing, judgement reasoning methods, natural language understanding and physical capabilities, such as mobility , navigation, gross motor skill, fine motor skills. It seems that the future, (AI) human invents machines which will have these human characteristics to do human similar behavioral job duties more easily and efficiently. It implies these above human occupations will be replaced by (AI) human invention machines in the future. Due to (AI) creation, it is possible to cause unemployment number of these above workers will increase because (AI) machines can do their similar job behavioral activities.

Consequently, employers won't need to employ many of these skillful labor. Otherwise, they can buy less number (AI) machines to attempt to do whose job activities more easily and efficiently. So, it seems (AI) machines will have more high work performance to replace these occupation workers' work performance. Finally, these occupation worker unemployment number will only increase when the (AI) machines had been invented to achieve to do their work behavioral activities absolutely success in the future.

3.3 Whether (A) technology machine labor
will replace human worker more or assist
human worker more

There is no single agreed definition of a robot how outcome of a task that is completed without human intervention. When some definitions require the task to be completed by a physical machine moves and respond to its environment, other definitions use the term robot in connection with tasks completed by software , without physical embodiment.

However, to answer the question : Whether (AI) technology machine labor will replace human worker more or assist human worker more. I shall indicate some examples to let readers to judge whether (AI) technology can create new jobs or reduce old jobs.

Firstly, I shall explain what (AI) function is. (AI) is a service robot that performs useful tasks for humans or equipment excluding industrial automation application . Thus, the classification of a robot into industrial robot or service robot is done according to its intended application. It is also a personal service robot or a service robot for personal used for a non commercial task, usually by lay persons . Examples are domestic servant robot, and pet exercising robot. It is also a professional service robot or a service robot for professional used for a commercial task, usually operated by a properly trained operator. Examples, are cleaning robot for public places, delivery robot in offices or hospitals, fire-fighting robot, rehabilitation robot and surgery robot in hospitals. Thus, these functions will be future (AI) application to our daily life necessaries or business necessaries.

However, some authors agree (AI) will bring negative outcomes of automation, due to raise competiveness, reduce human job nature. Otherwise, other authors argue (AI) will bring positive outcomes of automation, due to raise productivities, job creation, assist humans work.

On the positive outcome hand, robots can increase productivity . This is particularly important for small-to medium sized businesses both are in developed and developing countries economies. It also enables large companies to increase their competitiveness through faster product development and delivery. Increased use of robot is also enabling companies in high cost countries to re shore, or bring back to their domestic base parts of the supply chain that will have previously outsourced to sources of cheaper labor. Currently , the greater threat to employment is not a automation, but an inability to remain competitive. Automation has led overall to an increase in labor demand and positive impact on wages. The reason is that the middle-income/middle-skilled jobs have reduced as a proportion of overall contribution to employment and earnings leading to fears of increasing income inequality, the skills range within the middle income bracket is large. Thus, robots are driving an increase in demand for workers at the higher -skilled and with a positive impact on wages. This issue is how to enable middle-income earners in the lower-income range to unskilled or retain. Finally, the (AI) positive impact supporter who argue the future will be robots and humans can work together.

However, on the negative outcome hand, robots can substitute labor activities, but don't replace jobs. They believe that less than 10% of jobs are fully automatable. Increasingly , robots are used to complement and augment labor activities, the net impact on jobs and the quality of work is positive. Automation can provide the opportunity for humans to focus on higher-skilled, higher-quality and higher-paid tasks. Robots can improve productivity when they are applied to tasks that which perform more efficiently and to a higher and more consistent level of quality than humans. For example, increased productivity is enabling some firms, such as Whirlpool, Caterpillar and Ford Motors company in the US restructure their supply chains, bringing back parts of the manufacturing process to the country of origin. Thus, productivity gains due to robotics and automation are important not just at the company level, but also for build industry and nation competitiveness.

I suppose that productivity can be raised. What are the impacts of robots on employment? Firstly, the main focus of development has been on personal entertainment, which does not drive worker productivity (manufacturing production). When the internet (information and communication technology (ICT)) innovation. This is borne and by findings that manufacturing productivity, which has been driven by innovations in automation rather than consumer technologies, has government strongly than productivity in the services sectors of the economy in most nature economies. It seems (AI) automation will create many jobs in internet communication entertainment game industry. For example, many young people like to use internet to play any electronic games from computer or mobile at home or outside home conveniently. Thus, (AI) automation will increase demand to be invented to any new entertainment game from internet channel. It will need to employ many (AI) entertainment game inventors to create many automation entertainment games. Thus, (AI) automation in internet entertainment game industry will need human (AI) entertainment game inventors to invent the knowledge-based capital of (AI) automation entertainment games. The (AI) entertainment game inventors will need own research and development skills, form specific skills, organizational know-how skills, databased knowledge, design and various forms of intellectual property to do these (AI) automation entertainment game invention occupations in the future.

International Federation Of Robotics(2016) indicated that China will be as a major robotics manufacturer and user of robots, benefiting from jobs created by robot manufacturing and productivity gains from robot use. Chins had sold of robots to any one single market every year since 2017 year. The Chinese government has included a focus on robotics in its 10 year strategy. In order to achieve its target of a robot density of 150 units per 10, 000 workers by 2020 year. Thus, Chinese companies will have to install around 650,000 new industrial robots between 2016 to 2020 year, 2.5 times more than installed globally in 2015 year.

Hence, China (AI) manufacturing industry will need to employ many workers . It implies (AI) manufacturing industry will create many new occupations in China. Also, ministry of economy, trade and industry (2015) also showed that Japan currently has the largest stock of industrial robots in operations, primarily in the automation industry. Driven by a rapidly aging population and low productivity rates, the Japanese government has sights on a 20-fold increase in the use of robots in the non-manufacturing sector and a three-fold growth rate of labor productivity in the service sector both by 2020 year. Thus, it also implies Japan will need many robots to be provide to service industry. Due to robots will provide to serve any businessmen's clients. Thus, it is possible that the service workers won't be dismissed as well as it is depended on the serving job nature to decide whether Japan's service workers can still serve to their employer when the service (AI) robots are applied to whose employers.

Consequently, it seems that (AI) can create employment, Ministry of economy, trade and industry (2015) showed that such as China will develop the major (AI) automation manufacturing industry. The (AI) employers will need to employ many workers to manufacture any these different kinds of (AI) robots to satisfy China or overseas individual or business buyers needs. But, (AI) can also cause unemployment to the low skillful service workers. Such as if Japan some service businesses choose to buy any (AI) service robots to replace their service staffs to serve their clients. It is possible that the service staffs will be dismissed, due to (AI) robots can do such as their same service job duties to achieve better service performance. Thus, today, it is increasingly common for people to use robots in various situations at home and in retail stores, hotels and hospitals these service industries. Robots are classified into server types based on their functionality (service and utility robots or those designed to communicate with humans) and appearance (humanoid robots or mechanical robots). The type of robot, to which each country allocated particular importance in the advance of robotics, reflects the sense of values and preferences of its population. Thus, if the country has high population needs to use robots, then they will influence either more new jobs creation or more old job loss in the country's (AI) manufacturing or (AI) service industries both. For example, Japan respondents often associate the term " robot " with humanoid robots that can communicate with human and they have a high level of familiarity with robot. The US has the highest level of robot utilization at home and in retail stores with its people being the most enthusiastic about the future use of robots. Germany shows a strong tendency to consider robots for industrial purposes and its people feel strong effort to the presence of robots in their households.

In conclusion, to judge whether how (AI) will influence the country's employment to be better or worse. It will depend on the country home buyers (users) or business buyers (users) how to use (AI) for their daily needs. If the country , such as US retail stores need to use (AI) , it will have possible to reduce some or many retail service workers. Even, if the country , such as Japan has many home users need to use (AI) , it will not influence the employment market. Otherwise, it will raise (AI) salespeople numbers. Even, if the country, such as Germany and China will have many (AI) manufacturers, then it will create many (AI) manufacturing occupations for these (AI) manufactory workers.

Consequently, (AI) robots manufacturing and service needs will have positive or negative impact to any country's employment. It will depend on the (AI) service provision and service workers' job nature as well as the manufacturing workers of (AI) knowledge level to decide their employment chance in their country's employment market.

● Could work activities in China be automated
making in the nation with the world's largest automation potential?
Can (AI) technology influence China economy? Could China workers be affected and jobs made up of routine work activities and predictable? Will programmable tasks be particularly impact to China employment market ? When impact on labor market is likely to be gradual at the aggregate level, it can be sudden and dramatic at the level of specific work activities, rending some job obsolete fairly. Overall (AI) technology will raise digital skills when reducing demand for medium incomer inequality for China workers. It seems (AI) technology's effect on productivity could be crucial to China's future economic growth as the population ages are increasing.

In China, some biggest technological companies driving significant investments in research and development. Moreover, China is one of the leading global (AI) technology development county. However, China will need to focus on building its innovation capacity. For example, United States and United Kingdom are currently producing more influential (AI) technological research. However, if China planed to achieve (AI) technology success, it's traditional industries will need to develop technical know-how −to and overcoming implementation costs prepare to develop (AI) . When (AI) technology is introduced into China society, China government needs to raise concerning ethical, legal, technological security etc. business questions. Also, surrounding issues include privacy, discrimination, legal liability and regulation. It aims to encourage overseas investors to choose to invest (AI) technological industry to raise GDP growth and manufacturing industries income growth for long term in China. If China encouraged overseas (AI) technology investment in its country. It is possible to influence China employment market to be changed. Because (AI) technology will impact to influence China people daily life. Due to (AI) technology is introduced to China society, many rich people will prefer to spend to buy any high (AI) technological products for entertainment or learning or machine man driving etc. daily necessity activities. Then it will raise GDP growth and will raise (AI) manufacturers or related-(AI) technological manufacturers profit. It is beneficial to China because it can become one high knowledgeable and (AI) technological economical society. But it will bring bad influences to raise unemployment chance for the low skillful labor. In labor economy aspect influence , how (AI) technology can influence China low skillful labor unemployment ratio raising. The raising low skill labor unemployment reason is because China low skillful human labors are argued or are replaced by (AI) technology creating new challenges to introduce to influence China society of simply human manufacturing job nature to be changed to be high (AI) technology manufacturing job nature in any China factories. Moreover, when (AI) technology introduction to China, it will cause other related social challenges in China. The varied (AI) related challenges, including the difficulty of creating safe and reliable hardware for sensing and affecting (transportation and education), the challenges of gaining public trust, a low resource comities and public safety and security, the challenges of overcoming fears or marginalizing humans in China employment and workplace and the risk of diminishing interpersonal trust because the low skillful labors won't believe any China employers will give chance to employ them , due to (AI) technology will replace their skills and man manufacturing of productivity is much less to compare to (AI) technology manufacturing method.

● How does (AI) technology influence
the future of employment change?
Are future nature of jobs changed to computerization from (AI) technology? Where are the probability of computing occupations from (AI) technology influence? What is expected impacts of future computing on labor market from (AI) technology influence? John Maynard Keynes's frequently cited prediction of widespread technological unemployment " du to our discovery of means of economic the use of labor outrunning the pace of which we can find new used of labor" (Keynes, 1933, p.3).
In the future, (AI) technology will impact some nature of occupations to change computing. This chance will also influence some countries' economic change. For example, some factory human labors hand routine manufacturing tasks will be changed to computerization of routine manufacturing tasks by (AI) technological machine men hand

manufacturing method. it will cause a structured shift in the labor market, with workers reallocating their labor supply from middle-income manufacturing to low-income service occupations.

Arguably, this is because the manual tasks of service occupations are less computerization, as who require a higher degree of flexibility and physical adaptability. So, (AI) technology will influence the human hand labor skillful occupation nature of task cheaper , such as vehicle manufacturing , ship manufacturing, computer manufacturing, steel manufacturing, television, radio etc. home electronic products of heavy machine industry change. Due to (AI) technology machine man will be proper to be used to manufacturing these electronic products when the (AI) technology innovation can develop to the mature stage. Then, any countries manufacturers will choose to use (AI) technology machine man, instead of human hand production.

Supposing the future prices of computing are fallen, seriously, problem solving skills are becoming relatively productive, explaining the substantial employment growth in manufacturing occupations, involving cognitive tasks where skilled labor has a comparative advantage, as well as the increase education needs for (AI) technology computing of machine man subject study.

Prediction of education needs for (AI) technology student numbers will increase, due to manufacturing industry needs many (AI) technology students in future employment market. Another (AI) technology influence if the future (AI) technological innovation, e.g. machine man manufacturing or machine man service industries will both increase demand, then with more sophistic software technologies will be disrupted labor markets by marketing workers redundant.

For publishing industry, what is striking about the case in paper book publishing industry will be unpopular? Due to the electronic book publishing industry will be popular, e.g. Amazon publish . (AI) technology can influence paper book manufacturing method which is replaced by machine man electronic book manufacturing method as well as it will cause the computerization is no longer confined to routine manufacturing tasks. Due to (AI) machine man manufacturing technology will be proper to be used to manufacture any products in short time efficiently and effectively , e.g. electronic book products. In the future, if it is fact to occur this case, such as (AI) technological machine man manufacturing method will be adopted (applied) to manufacture electronic books or any products in possible. (AI) technology will cause many manufacturing workers are unemployed. It is beneficial to employers, who can reduce to spend much wages expenditure to employ manufacturing workers, but it will cause many manufacturing workers loss jobs and reduce income to support whose families lives. It will cause social challenges, e.g. increasing stealing crimes if the manufacturing workers had not other skills to find other jobs to do easily. So, manufacturers need to concern over technological unemployment which will be hardly future phenomenon if who decided to dismiss all manufacturing workers, due to (AI) technology machine men replace to them.

If (AI) technology can be innovated to produce any kinds of machine man to serve any service or manufacturing industries successfully. Then, it will bring these questions: Can future that workers be influenced to be automation employment and productivity by (AI) technology influence? Does it impact to influence the (AI) technology countries' productivity and growth and natural resources development and labor markets and evolution of global financial markets and economic impact of technology and innovation and urbanization etc. issues? How will automation transform the workplace? What will be the implication for employment? What is likely to be its impact both on productivity in the global economy and on employment?

In fact, automatic of activities can enable businesses to improve performance by reducing errors chance and improving quality and speed, and same cases achieving outcomes that go beyond human capabilities. Some economists indicate (AI) technology would give a needed boost to economic growth and prosperity have of the working age population in many countries. Based on the scenario modeling, they estimate automation could raise productivity growth globally by 0.8 to 1.4 % annually. They also indicated that almost half the activities people are almost $1.6 trillion in wages to do in the global economy have the potential to be automated adapting current demonstrates technology, according to their analysis of more than 2,000 work activities across 800 occupations. When less than 5% of all occupations can be automated entirely using demonstrated technology, about 60% of all occupations have at least 30% of worker made activities, that would be automated. More occupation will change to be automated. They also indicated for business performance benefits of automation are relatively clear, but the issues are more complicated by policy making to attract foreign investors. Beyond technical feasibility, the cost of technology, competition labor will include skills and supply and demand dynamics, performance benefits and beyond labor cost savings and social and regulatory acceptance will affect the automation. Their predictions suggest that half of today work activities could be automated by 2055 year, but this could happen 10 to 20 years earlier or latter depending on the various factors in addition to their wider economic condition.

Some scientists suggest (AI) technology is finally starting to deliver real-life business benefits. Computer power is growing significantly , algorithms are becoming more sophisticated and perhaps most important of all, the world is generating vast quantities of the fuel that powers (AI) technology data billions of gigabytes of it every day. Also, online firms are digital natives, such as Google online search service company is investing on (AI) technology. For new though most of the news if coming from the suppliers of (AI) technologies. And many new users are only in the experimental phase. Few products are on the market or are likely to arrive these soon to drive immediate and widespread adoption. As a result, analysts believe (AI) technology's potential will give true economic benefit in the future. (AI) industry will introduce to suppliers and users to raise economic potential of (AI) technology.

In the future, (AI) technology systems can solve business problems. Some scientists categorized those into five technology systems that are key areas of (AI) technology development: robotics and autonomous vehicles, computer vision language virtual agents and machine learning , which is based on algorithms that learn from data without replying on rules-based programming in order to draw conclusions or direct an action.

Such as computer vision and language includes natural language processing, analytics, speech recognition technology, some are about learning from information, such as about machine learning and others are related to acting on information, such as robotics, autonomous vehicles and virtual agents, which are computer programs that can converse with humans. Machine learning and a subfield called deep learning are artificial intelligence applications.

● Can artificial intelligence impact
global economy growth?

Artificial intelligence (AI) is a term first defined in 1956 year. It is a branch of computer science that aims to create intelligent machines that work and react like humans. In contrast today, 60 years later, (AI) is characterized by a number of applications, including computers playing games against humans and understanding human languages, virtual personal assistants, and robotics which involve computers seeing , hearing and reacting to sensory stimuli. In the future, technologists predict for (AI) technology ranging from (AI) being used as a tool to aid relatively simple processes for robots with human like mental capabilities, who expect (AI) technology can emulate human performance by learning, coming to mind its own conclusions, understanding complex content, engaging in dialog with people, enhancing human cognitive performance or replacing humans in executing both routine and non-routine tasks. In existing industry, (AI) technology is used , such as targeted advertising and virtual used personal assistant as well as the (AI) technology that my exist in the future, such as robots with human vehicle processing capabilities.

The range of (AI) technology's progress in the future will determine the economic impact future of (AI) technology on the global economy with more limited advances and applications (i.e. weak (AI) only) corresponding to more limited economic impacts and more substantial progress, i.e. strong (AI) technology is corresponding to more significant economic impact.

(AI) technology learning that automates analytical model, including predicting cause-and-effect relationship from biological data, identifying new drugs, self-driving cars and protecting against fraud etc. functions. Also (AI) learning can improve natural language processing that allows computers to continue to better analyze, understand and generate language to interface with human using the natural human language, virtual personal assistant, helps users by providing scheduling appointment, reminds organizing personal finance and finding providers of various services, machine vision allows (AI) machine man to identify object, scenes and activities in detect pedestrians and bicyclists.

We expect the economic effects of (AI) technology to include both direct GDP growth from sectors that develop or manufacture (AI) technology and indirect GDP growth through increased productivity in existing sectors that employ some from of (AI) technology. If (AI) producing sectors could grow, then it could lead to increase revenues and employment of (AI) technological professionals within these existing firms as well as the potential creation of entirely new economic activities to any countries' societies productivity improvement in existing sectors could be realized through faster and move efficient processes and decision making as well as increased (AI) technological knowledge and access to information available in societies easily.

In the future, if (AI) technology is an increasingly critical component of more products, it will become an integral part of necessary products of many people's lives. The extent of (AI)'s economy effort is also likely to vary from region to region, thought variation may be more dependent on the predominate economic activity of a region and the (AI) ability can influence economic activity, rather then the economic or developmental status of the regions. (AI) technology can move accessibility and can use source development to do international business between one country and another country.

So (AI) technology has the potential to give benefits to different income chooses and to bring significant gains to both developed and developing countries. For agricultural technology, (AI) has the potential to optimize food production around the world by analyzing agricultural regions and identifying what is necessary to improve crop yield. In total, (AI) technology gives greater economic impact to any countries agricultural regions if which implemented (AI) technology to grow crop , fruit etc. food production in the farms.

Investment in (AI) technology is such as capital investment to any countries' public or private enterprises. So, it will have large economic impact to the future . If the (AI) technology is reasonable invested to the different needs aspect by the public or private enterprises in the country. Then, it will have good economic impact to the country in the future. However, when (AI) technology is likely to affect both the productivity and employment components of economic growth in many sectors. Significant public debate has focused on projections of (AI)'s effect on the labor force. However, for instance, some researchers have argued that the rise of (AI) technology and automation will led to significant unemployment as capital is substituted for the low skillful labor. So, they point to the concern that the increasing sophistication of (AI) technology may balance skilled and semi-skilled workers and the reduce the size of the middle class. However, this is not a new argument, due to (AI) technology negatively affecting the labor force and leading to mass unemployment. Because the (AI) technology is the substitution of machinery for human labor. Although, employment in certain industries, has been reduced in the past due to technological advancement. For long term, the labor market has adapted to the introduction of new technology, giving rise to new jobs in new areas. (AI) technology may also be accomplished without a reduction to total employment in the long-term to some Asia countries, such as Hong Kong and Japan. Because Hong Kong and Japan many low skilled labor, e.g. security, cleaner who complaint that employers need them to work long time hours. (abnormal working hours) e.g. one day 12 to 15 working hour per day. Hence, if (AI) machine means invention technology success. Security or cleaning job can be worked from (AI) machine man in some hours every day in order to reduce the long time working hours cleaners or security workers, e.g. one (AI) machine man works 4 hours for cleaning or security job, one day as well as another cleaner or security labor only needs to work 8 hours one day. So total security or cleaning employers can employ 12 hours machine cleaners or security workers and human cleaners or security workers in one day. For long term benefit, Hong Kong or Japan every security or cleaning worker does not need to work 12 hours minimum working hours one day. They won't feel tried and bore and without private with whose families, so who will accept to do these cleaning or security jobs, even they can raise work efficient and performance when who feel happy and health.

So, (AI) technology of machine man invention can raise low skillful labor efficiency and it can help them to avoid abnormal working hours demand in some busy work life countries, such as Hong Kong and Japan. Before, one Japan female labor feel unhappy to work, due to who often needs to work abnormal working hours for her employer and who has less sleeping and without any private time to enjoy her life with her families every day. So this abnormal working hours factor causes her to do commit suicide behavior, then she is die unlucky. So (AI) technology of machine man invention ought avoid abnormal working hours demand for employer in any countries in the future.

The most important occurrence to any employers, some researchers had attempted to do one experiment to find that private research and development , venture capital and public research and development investment all have strong net effect or economic growth with venture capital funding further having the strongest such effect from (AI) technology. The researchers hypothesize the venture capital investment contributes to economic growth through (AI) technology innovation and by the capacity of an economy to use existing (AI) technology knowledge to increase productivity. They predict the impacts of venture capital, business-research and development and public research and development can raise multi factor productivity from (AI) technology introduction.

Can (AI) technology influence the economic development to developing countries? The developing regions of the world contain most of natural resources. If one day, (AI) technology has invent one kind of machine man which can assist any gas or oil workers to seek any new oil/gas natural resource locations easily. I believe that (AI) technology can help these natural resource exploitation countries will gain economic benefit more easily. So, (AI) driven technology can be used to change to create any new opportunities to address poor management or resources and improve human well being, such as Africa Latin America and India can use (AI) technology machine man to seek any oil/gas natural resource countries exploitation activities to attempt to gain much economic benefits.

● Why will (AI) technology grow economic
development ?

Nowadays, increases in capital and labor are no longer driving the levels of economic growth, such as (AI) technology. The ability of increase in capital investment and in labor of traditional drivers of production, have no longer to be enjoyed in most developed economies ,e.g. developed country, US, UK . However, artificial intelligence has the potential to overcome the physical limitation of capital and labor to avoid missing out on this opportunity. So, policy makers and business leaders must prepare for and work toward a future with artificial intelligence. They must do with the idea that (AI) is another simply method to enhance productivity method . Rather they must see (AI) as the tool that can transform thinking about how growth is created.

Economists have always thought of new technologies are as driving growth their ability to enhancing. It can replace labor and capital factor of production. So, it brings this question: What is the factor of production (AI) technology characteristics. They key factor is to see (AI) technology as a capital-labor .

(AI) can replicate labor activities at much greater scale and speed, and to even perform some tasks began the capabilities of human. For example, by using virtual assistants , 1000 legal documents can be reviewed in a matter of days instead of taking three people six moths to complete. Some (AI) technology may be one kind of factor of production in the future. For another example, people will work in workplace digitalization environment. So, in the future, working environment and information management are automated. Such as Konica camera sale company will use workplace digitalization. So , (AI) technology can provide workplace digitalization in order to raise productivity efficiency. (AI) technology will be one kind of production which is replaced by workplace digitalization and it will grow any organization productivity efficiently. Then, (AI) technology will assist overall social economy growth , due to productivity is raised and products can be produced in short time to prepare to sell in consumption market. So, time will be shortened to increase GDP growth fast for the development of (AI) technology countries.

● How can (AI) technology impact to global economic and social and psychological changes?

What will be the development of (AI) technology and predictions concerning the future evolution? The computers and robots will develop conscious, intelligent and minds into humans, enhancing psychological and behavioral abilities and allowing for direct communication with (AI) minds. (AI) technology will be impacted human life by (AI) technology information communicative and environmental influence. A " world brain" and " world mind", this psychological system will be enhanced and enriched the capacities of both individual and collective cognition by (AI) technology of service industries.

(AI) technology with influence these human needs of service industries changes, such as , biological science, finance, entertainment, business, biological science, transportation, communication military etc. The personal computer evolution, the internet and the world wide web which exploded on the scene, linking business, homes, schools, social organizations which were a completely unpredicted phenomenon to influence human life. Kurzweil (1999) predicts that by 2029 year, most human communication will be with machines. According to Person, by 2100 year, there will be human machine convergence.

How can (AI) technology influence environmental protection to make benefits to farming economic growth? (AI) technology can be applied to predict how to solve environmental pollution challenge to avoid to damage any crop or vegetable or rice or fruit etc. food growth. Because environmental experts can gather global environmental pollution data from an environmental database to build a perform a systematic analysis from (AI) technology. The first step is this broad analysis can include understanding, statistical and data gathering techniques to obtain the relevant data, the correlation among the variables involved, and a list of possible models. The next step is to select a set of methods and models that cover all kinds of knowledge and functionalities needed for the decision making process. Once the models are selected, they must be fully implemented by means of machine learning , data mining, statistical or numerical technique. After that, those models must be integrated to build the whole EDSS. The EDSS must be tested to check its performance, accuracy, usefulness and reliability, both from the user's and (AI) technology/computer scientist's point of view. If these is any wrong feature in any development stage, such as model's integration, models' implementation, selection of models, database, problem analysis etc. the developers must come back in the update the required components. When the evaluation phase is all right, the EDSS is ready to be applied to the environment. The great contribution of artificial intelligence to EDSS the integration of several methods complementing the classical statistical models/simulation , statistical analysis, linear models, etc. and numerical models (control algorithms, optimization techniques etc.) .

This cooperation makes the resulting systems more reliable and powerful in coping with real world environment systems. Date interpretation has been a principal area of research in (AI) technology since the very beginning. The most demanding problem in the environmental assessment context. Knowledge representation permits the definition of the different types of data that the existing methods adapt to the process. There is also a lot of work to clean, repair and transform the huge available quantities of raw data. Apart from this, the availability of meta-information or background knowledge is required to guide the process. Data mining is multi-disciplinary: It covers expert systems, data based technology, statistics, data visualization and unsupervised machine learning. These techniques operate at the level of data and background information, where numerous and often incompatible new commensurate pieces of information from disparate sources have to be brought together (K, Fedra, 1994).

So, it seems that in the future, (AI) technology with the increasing maturity in particular those related to knowledge and engineering, new dimensions can be assisted to users in environmental decision making are available. For example, many environmental systems are characterized both by incomplete models and by limited data. Hence, in the future, (AI) technology will be applied to predict climate change to reduce crop or fruit etc. food agriculture challenge by climate change bad influence.

● Will (AI) technology influence digital economy change to manufacturing industry ?

To understand how the manufacturing business must adapt to prosper in the technology, we need to understand how (AI) technology will change us to shape our daily habits to satisfy our expectation of products to how we shop and even the immediate of the entire process. For example, taxi services are in the crosshairs as on demand transportation services like, available of the touch of a smart phone button expand. In fact, Yellow lab, US country , san Francisco city's largest taxi company is filing for bankruptcy as the industry starts to change faster than almost anyone expected. However, at this point, its more than an app that is changing, some our taxi passengers renting taxi transportation to catch consumption behavior.

(AI) technology will influence digital economy for taxi passenger's individual customer experience, offering a growing renting taxi to catch of service and feedback opportunities when any one taxi passenger who chooses to use mobile phone app online tool to prepaid to rent any taxi more easily.

Also in the long term, (AI) technology can influence vehicles drive themselves of behavior. Already, companies like Google and GM are working on projects to bring fleets of autonomous vehicles to cities at the path of a button.

Moreover, this on-demand service model is beginning to appear across a much broader range of markets. For example , Amazon company is investing in its own fleet of trucks, planes and even drone at the same time as it pushes for same-day delivery of products. As some point, vehicles will be autonomous too. So, it seems that (AI) technique will influence any transportations choose to use digital autonomous driving technology in the future . For Amazon company case, it is not stopping of logistics. It is also aiming to automatically manage the supply of consumer home products with its recently launched Amazon replenishment service, Dash. Dash is a digital service that enables that connected derive to automatically order physical products from Amazon when supplies are running low. So, it seems (AI) technology will be applied to logistic function by digital technology method introduction in the future.

Hence autonomous vehicles will optimize industry supply chains and logistics operations through increased efficiency and flexibility. In fact, fully automated and lean supply chains will keep reduce load sizes and inventory by leveraging smart distribution technologies and smaller autonomous vehicles by machine man assistance. If Amazon continues to grow market share for online sales by reducing effort required by the consumer to place an order, when also contributing the almost immediate delivery of products to the doorstep. So, it will further fuel the trend toward on-demand derive. As Amazon company fuels the on-demand economy, consumers will expect immediacy in more parts of the digital economy. On top of speed, consumers increasing expect more personalization options.

So, (AI) technology will influence digital manufacturing, such as Amazon publishing to monitor every aspect of every process in real -time and communicating to self-optimized deep learning robotics, new methods of high volume and high customization will become possible. Then, as products merge into product platforms and even services, manufacturers have the opportunity to provide components and platforms used by smaller players. So, (AI) technology will influence manufacturing industry to choose automated SMI lines, robots installed, automation engineers.

Another future (AI) technology development can be applied to space science aspect, such as Automation engineering space in manufacturing process to achieve digital manufacturing benefits to any businesses in the future. Such as reducing cost, shortening manufacturing time, raising efficiency, shortening delivery products to client individual time. How can artificial intelligence give the need and advanced fast and evaluation methods benefits for space exploration? When US NASA (space exploration organization) achieves any space exploration missions, it will answer this question:

When is it useful to have a machine use (AI) technology to achieve a decision? After all, after millions of years of space exploration and rough 10,000 years of civilization, humans are usually quite good at making decisions in complex uncertain environments. Through, Johns Hoplains University's Applied Physical Lab. Research in (AI) technology enabled systems, which has identified three general use cases for (AI) technology to explore space mission:

First, for some tasks (AI) technology is more cost effectiveness than human. Second, (AI) technology is better suited than humans at solving some, but not all problems. Third, (AI) technology allows NASA organization's space exploration mission to develop machines that ate capable of responding faster than when a human is in the decision loop (D. Scheidt, 2012, A. Castano et. al. 2008).

So, the use of (AI) technology to enable science by observing the pace of rapidly evolving phenomena was demonstrated. It is more effectively coordinating and (AI) technology utilizing to earn economic benefits to use for space exploration mission.

However, (AI) technology also have current risk for space exploration. Today (AI) technology is immature and requires further development to reach its potential. For instance, the (AI) technology algorithms that detected the dust derive could not have identified whether the Martain weather represented a threat to the cover. Also it can not yet use instrument input to determine what, where and how to autonomously make the next space science measurement. An equally important factor limiting (AI)'s deployment is that lacks the methodology and technology to effectively test (AI) technology. So, the challenge will testing (AI) enabled system is how (AI) performance can be measured. It would be NASA organization's difficulty to find (AI) technology to develop to carry on researching any space exploration missions in the future. However, (AI) technology will be a good economic benefit choice for space exploration mission in the future.

● What is artificial intelligence potential
benefits and ethical considerations?
The ability of (AI) technology systems to transform vast amounts of complex information into insight has the potential to help solve manufacturing or service challenges for human needs. However, to reap the societal benefits of (AI) systems, humans will need to trust then and make sure that which follow the same ethical principles, moral values, professional codes and social norms that we humans would follow in the same scenario, research and educational efforts as well as carefully designed regulation in order to achieve the most effort of economic benefits goals. For example, international business machines corporation (IBM) is actively engaged both competitors , in global discussions about how to make (AI) ethical and as beneficial as possible for people as social economic benefits.
(AI) is usually defined as the " capability of a computer program to perform tasks or reasoning processes " that human usually associate to intelligence in a human being. Often, it has to do with the ability to make a good decision, even when there is uncertainty, too much information to handle. As an example, play chess or complex card games of entertainment activities is believed to need some form of intelligence in a human being, as well as

choosing the best medical facilities in a difficult medical case, or creating something new, such as mathematical theorem or even some form of act, or even driving automatic machine man (self driving vehicle) replacing human driving in the middle of a crowded city.

(AI) needs depends on what we consider being intelligence in the behavior of a human being act a certain point in time. If human belief about human intelligence changes and we don't believe any longer that a certain task requires intelligence, then a computer program performing that task is no longer part of (AI), it becomes just another boring computer program. So, it means that (AI) technology will replace some old computer programs, if human can invent new generation of (AI) software for any functions or activities to satisfy human needs.

As IBM, it argues intelligence. This means that we aim to build systems that enhance and scale human expertise and skills rather than replacing them. We therefore focus on practical applications of (AI) capabilities that assist people in performing well-defined tasks of needs by exploiting and wide range of (AI)-based services. We also use the term " cognitive computing" it is mean a comprehensive net of capabilities based on technology. It comprises the fields of machine learning, reasoning and decision technologies, language, speech and vision recognition and processing technologies, high performance and high efficient functions for any industries or individual consumers needs. For example, robotics, which are usually very good at doing what which are supposed to in any environment, much have public shopping center, factory etc. places which need simply services from the robot (machine man), such as cleans the floor of our houses to the robot that can work together with humans in production chains, passing through the warehouse, robots can take care of the tasks of an entire warehouse and the companion robots like Nao, Pepper, Aibo and Giraff, who can entertain use, talk to use and help elderly people to stay connected to their friends, relatives and doctors.

Google company is building automatic machine (self-driving cars) and has acquired more than 10 robotics companies. Facebook had opened whole new research facility only on (AI) research. Apply computer has developed Siri. Microsoft computer company has built a similar personalized assistant. Google has Deep mind, a UK company whose long term aim is to build general (AI) and has already great potential to win game to the world champion and IBM is investing a huge amount of resources in applying its Watson cognitive computing system to the medical domains to finance and to personalized education. In Europe, IBM is establishing new centers in Munich and Milan focused in the application of cognitive computer capabilities to the internet of things and healthcare respectively.

For example, automatic machine man (self-driving cars) are all about (AI), which used to be able to see what happens in the street (signals ,lanes, other cars, pedestrians, traffic lights, which need to able predict what other cars and pedestrians will do, and who need to be able to cope with unforeseen situations. Since, most car accidents are due to human fault, it is estimated that the adoption of self-driving cars will save about half of the lives that are usually last in car accidents.

IBM Watson company has to understand spoken language, make sense of massive amount to text , respond correctly to questions in many categories, as well as assess its own confidence in responding to such questions. In the future, (AI) technology can own question/answering capabilities that would be very useful, for example, in assisting a doctor when trying to some to the correct diagnosis for a patient and to propose the best therapy .

Intelligent machines can also rely on huge amounts of data to be used to learn how to make better decisions. This data comes from all of us over the years Facebook users have uploaded more than 250 billion pictures and every day who upload about 350 million more. Every second, we submit 40,000 google search queries. So, (AI) technology will be connected through the web from appliances to traffic lights from cars to watches. Other tasks that are very easy for humans are physical and manipulation tasks, such as walking , running, picking up an object to make its shape and location, restricted environment. But (AI) machine man technology still not able to have the general physical and manipulation capabilities even of a 6 year old.

So, it brings this question: Why do (AI) scientists need to concern ethics? Because (AI) technology is complex, information into insight has the potential to reveal long held secrets and help solve some of the world's most difficult problems. (AI) systems can potentially be used to help discover insights to treat disease, predict the whether, and manage the global economy. So, ethic issues is important to and (AI) scientists . If any one new (AI) technology research investigation could success, it will be a secret to and the (AI) scientists can not permit to their loyalty to any competitors to damage the fair (AI) technology products trading market. The country (countries) (AI) technology scientists need to concern ethic issues, who need to keep secrets for their countries economic or/and social benefits.

This is moral issues to any countries/country loyalty is whose countries intangible assets. They can not sell (AI) loyalty to any their countries to assist whose economic benefits immorally.

● How can (AI) technology influence to global health care economy development?

According to (AI) lecturer analysis, when combined key clinical health (AI) application can potentially create $150 billion in annual savings for the US healthcare economy by 2026 year. (AI) technology is re-winning modern conception of healthcare delivery. It enables machines to sense, comprehend, act and learn. So which can perform administrative and clinical healthcare functions (Accenture, 2017).

It will help health care service organizations to reduce health care cost, will improve and raise service quality and access. So, (AI) health market size will be predicted growth. (AI) applications in health care include robot-assisted surgery, virtual nursing assistant, administrative workflow assistant, fraud detection, error reduction connected machines, clinical trial participant identifier, preliminary diagnosis, automated image diagnosis and cybersecurity.

What kind of benefits (AI) technology can contribute to healthcare service? (AI) technology can deliver what many health care organizations need, such as financial and operational of labor costs, digital expectations from patient consumers how to use (AI) technology to solve interoperability challenges in any healthcare organizations. Also (AI) technology can be applied to wellness an d lifestyle management, diagnostics, delivers financially but also way of organizational and workflow improvement. So, (AI) technology will be continue to become most prevalent and adoption to healthcare organizations , which must need to enhance structure to be position to take full advantages of new (AI) technological capabilities. (AI) technology can change the nature of work and employment is rapidly changing to make the best use of both humans and (AI) talent in healthcare industry in the future. For example, (AI) technology offers a way to fill in gaps and the rising labor shortage in healthcare. According to Accenture analysis, the physicians shortage is increasing. However, (AI) technology will manufacture healthcare machine men to replace physicians in future one day(2017). Hence, (AI) technology will be invented to raise health care service staffs work efficiency and performance in any hospitals or clinics in the future.

In conclusion, (AI) technology will raise efficiency for any service or manufacturing industries in the future, although, it is possible that it will also rise low skillful workers unemployment numbers. But, the most important influence to human technological innovation will be risen and it will influence human life will be changed to be better, e.g. self drive cars, health care physician machine men, machine man cleaners etc. intelligent machine men will be manufactured to serve for our daily life. Furthermore, (AI) technological products will influence countries trading, some low technological development countries manufacturing businessmen can choose to buy any (AI) products to raise whose productivity and efficiency and reducing cost to achieve economic cost saving result. Also, GDP of trading growth income will increase to the (AI) products sale countries. Hence, it will be beneficial to economic development to both developed and developing countries both in the future as well as (AI) scientists time and money spending will be valued to continue to invest (AI) technology development for human life and economy benefits for long term.

In conclusion (AI) technology will raise macro economy growth and it can create many (AI) jobs , but it also raise the low level technological worker unemployment change. In the future, (AI) technology can be applied to digital technology to attempt to invent any new undiscovered (AI) and digital technology. So, it needs any scientists to continue to research how digital and (AI) technology can be mixed to satisfy human's future undiscovered needs.

Reference

A. Castano et. al. " Automatic detection of dust devils and clouds at Mars" Machine vision and applications, Oct. 2008, vol. 19, no 5-6, pp. 467-482.

Accenture, " Why artificial intelligence is the future of growth"(2017) <http://www.accenture.com/us-en/insight-a   rtificial-intelligence-future-growth>.

D. Schedidt , Unmanned Air Vehicle Command And Control, Handbook Of Unmanned Air Vehicles, Springer-Verlag, 2014. Facebook (AI) Research Available at https://research.facebook.com/ai, research at google, machine intelligence available at

http://research.google.com/pubs/machineintellige nce.html; micro soft research-machine learning and artificial intelligence available at http://research.microsoft.com/en-us/research-   areas/machine-learning-ai.aspx.

International Federation Of Robotics, 2016. IFR press release world robotics report. IFR, org . 29 Sept. Accessed Feb. 01, 2017.

http://www.ifr.org/news/ifr-press-release/world-robitics   report
-2016-8321.

K, Fedra , "GIS and environmental modelling" in environmental modelling with GIS, edited by M.F. Goodchild.B.O. Parks and L.T. Steyaert, Oxford University press, pp. 35-50, 1994.

Keynes, J.M. (1933). Economic possibilities for our
grandchildren (1930). Essays in persuasion, pp.358-73.

Mckinsey & Company (2013, May). Disruptive
technologies: Advices that will transform life,
business and the global economy , USA.

Ministry of economy, trade and industry, Japan, 2015, Japan's robot strategy. Ministry of economy, trade and industry.

Ray Kurzweil , The age of spiritual machines (1999) is cited numerously through this chapter: Kurzweilai.net http://www.kurzweilai.net

Rich, Elaine & Knight, Kevin, Artificial Intelligence Second Edition, 1991, New York; Mc-Graw-Hill.

CHAPTER II

How robots raise student learning effort

Nowadays, artificial intelligence (AI) is widely knowledge to be one kind of the dramatic technology. However, it is expected to continue, to have a disruptive impact on human's private and public life, so defense and security will be no exception. But how exactly will these be affected ? How will (AI) defense and security is incremental in nature? If (AI) technological machine men are applied to teach students in education aspect, is it better to my next generation learning development more than they are applied to war attack aspect.

To research why artificial intelligence (AI) has possible to be used to cause autonomous weapons by human. We need to understand these three aspects of relationship. They include cybersecurity and artificial intelligence and machine learning and autonomous weapon systems relationship between of them. Basic on (AI) machine can be invented to learn any new knowledge, so if (AI) machine men are taught how to attack enemy, which will be such as human soldier function. But, if (AI) machine men are taught how to learn university knowledge to teach students. Then, they will be such as human lecturer function. So, when (AI) is invented to own human mind and judgement and learning abilities, then they will be either human's enemy or human's assistant, such as university lecturer's assistant.

Firstly, we need to know what is the mean of artificial intelligence and cyber defense/offense? It means defense of critical networks: real time, pattern finding, anomaly seeking, it must utilize machine (AI) learning algorithms to efficiently, and instantaneously respond to potential network threats as well as it means human on or out of the loop. On the loop : it means anomaly detection: human notified, IT analysis, response. Out of the loop: it means anomaly detection: (AI) decides best method of response: quarantine, honey pot monitoring, hack-back. Thus, it is possible that (AI) can be used , such as autonomous cyber weapon. If (AI) is applied to make the decision best method of response to learning aspect, such as univeristy different subject knowledge. Then, it will be one good technological educational tool to teach university students.

In simplicity, (AI) can be one of scientific weapons platform or one of university teaching tool. When one day, it is invented to be applied to control war planes to fly to any countries to attack enemies or it is invented to be seemed to human to replace soldiers to bring guns or any weapons go to other countries to attack. So, it is possible that future any war defense planes, (AI) technological automatic control weapon can be replaced of human soldiers or war plane pilots to control any war defense planes to go to different enemy countries to attack them easily. It is very horror matter to threaten global human's ourselves life in the future , if (AI) automatic control war defense planes or (AI) automatic control machine soldiers were invented successfully. Otherwise, when one day, (AI) machine lecturer is invented to be applied to learn university different subjects knowledge to replace lecturers to copy lecturer's every prepared lecturer course to speak to let students to listen when they are sitting in university halls as well as the (AI) machine lecturer can make analysis and judgement response to answer every student's enquire immediately after it had speaking all courses to students to listen in lecturer hall every time. Then, it can let human lecturer does any education job duty, e.g. research education work. So, (AI) machine lecturer will be future human lecturer's assistant in future one day.

Finally, the most serious (AI) technological invention risks are human is unknown these aspects of (AI) absolutely: They are not simple automatic systems, learning reasoning, communication of " self-aware" systems. Thus, human will face (AI) technological invention risks or threats if human invent (AI) machine man to learn how to attack enemy. Otherwise, if human invent (AI) machine man to learn how to teach university student. I believe that my future university students can raise learn ability and writing ability and reading ability from (AI) machine lecturer teaching more than human lecturer teaching.

1.1 Online technology and online book
technology influences artificial intelligence
mind development

Nowadays, online technological invention bring online book technological development. Also, artificial intelligent technological machine men had been invented to link internet to do any jobs, e.g. children can find any data from artificial intelligent machine men when the artificial intelligent machine man had been installed internet and computer function, then children can find any online books to read from the artificial intelligent machine man. Such as Japan artificial intellgent machine men had installed computer and internet function, the Japan family children can find any online books to read from the artificial intelligent machine man at Japan any families' homes conveniently. Hence, it implies that future one day, artificial intelligent machine has possible to be invented to own human's reading and/or writing abilities.

For example, online book publishing is one kind of popular internet technology. For example, Amazon publish is as a business model with many potential advantages, relative to a physical operation. It held out the potential of lower book inventing and distribution costs and reduced overhead. Consumers could find the books, they were looking for more easily and a variety book topic choices could be offered for sale. It can accept and fulfill orders from almost any domestic location with equal ease. And most purchasers made on its site would be exempt from sales tax. One Amazon strategy hand, it would have to make its returns and redress processes transparent and reliable, and offer other ways for clients to learn, as much about the book possible before buying. Future online book market development trend, such as Amazon, Barnes & Noble etc. online book shops.

Hence, online book store technology can be applied to artificial intelligent technology. Such as artificial intelligent machine men can apply computer technology to learn the abilities of reading and/or writing any books either on paper or on computer. Hence, it is possible that artificial intelligent machine men will have similar human's writing and/or reading books ability when they own human's mind ability. However, it bring this questions: Can artificial intelligent machine men own human's mind abilities? If they own human's mind abilities, is it mean that they can write and/or read any books? Can artificial intelligent machine men own human's mind abilities to create to write any books? Can artificial intelligent machine men own human's mind abilities to read and make any judgements or decisions more accurate than human's judgements or decisions? To answer these questions? I shall indicate that online book reading and writing technology can be applied to artificial intelligent machine men reading and writing technology. Because they are similiar computer mind technological development. So, I believe that future artificial intelligence machine men can be invented to own similar human's reading and writing's mind abilities in future one day.

I believe artificial intelligence and online technological reading abilities are very similiar. Nowadays, computer can be invented to attempt to read and write any books by human. Why can not artificial intelligent machine men replace computer to read and write any books? Artificial intelligent machine men can replace human to attempt to write or/and read books, due to artificial intelligent machine men had invented to own human mind to do some jobs and their mind had been invented to be similiar to human behavioral abilities to do these behaviors, e.g. cooking, driving, playing games, singing songs, speaking, listening, frighting etc. different human's abilities. So, it seems that artificial intelligent will be possible to be invented to own human's mind abilities to do any writing or reading behaviors or functions.

1.2 Prediction of artificial intelligence
reading and writing abilities

development

What is future trend of artificial intelligence reading and writing abilities development? To answer this question, we need to know what benefits of artificial intelligent machine men can attribute to human's needs when they can own any human's mind to read or/and write any books.

I shall indicate e-books reading and writing example, if artificial intelligent machine men can be invented to own human's mind to write and/or read e-books on computer. Then, it brings this question: Can artificial intelligent machine men assist human to learn to do judgement to solve any challenges?

I believe that when artificial intelligent machine men can be invented to own human mind to write or/and read any books, then they will own human's mind ability to make judgement to solve any challenges more accurately, even their decisions can be more accurate to compare to human's decisions. So, artificial intelligent machine mens' writing and reading ability is the main factor to cause their mind to do any judgement in order to make any decisions more accurately. Consequently, in future one day, artificial intelligent machine mens' writing and reading ability will be invented to similar human's reading and writing abilities as well as their minds can also be invented to similar human's minds as well as their judgement abilities can be invented to similar to human's judgement abilities to make any decisions more accurate.

1.3 The influences when AI is invented to own human's mind and judgement abilities

Finally, I shall discuss what are the influences when AI is invented to own human's mind and judgement abilities in our future job market. The achievement of artificial intelligent (AI) machine men achievement requirement of owning human's mind and judgement abilities which requires extensive manual labor, and by augmenting the calling process with machine learning, the process where speed and accuracy are needed to close to human's mind and judgement abilities. Expert human race callers now have better information at artificial intelligent machine men at their fingertips faster.

Hence, if the above those requirements are achieved to satisfy artificial intelligent machine men ind and judgement abilities demand to close or exceed humans' mind and judgement abilities. Then, I believe that future human's some simple jobs must be replaced by (AI) machine men. Even, human's some professonal jobs, e.g. lawyer, accountant, administator, typing etc. professional skillful jobs, which will be either replaced or will be assisted by (AI) machine men. For example, (AI) machine men learn how to type english or other language words to do typing job ; they can learn how to apply accounting knowledge to record any firm's income and expenditure record of accounting job; they can also learn how to assist architects to design any architectural building drawing plans to do architect jobs; they can learn how to analyze any court evidences to judge any criminal or civil cases and assist lawyers to give legal advices to achieve more reasonable judgement for any legal cases; they can also learn how to assist firm's managers or administrators to manage any organization teams efficiently.

Consequently, when (AI) machine men can be invented to achieve to exceed human's mind and judgement abilities level. Then, I believe that they can do instead of human' simple jobs, which can do even human's more difficult and more judgement brent requirement of professional skillful jobs. So, (AI) machine men must need to achieve to do any jobs, they are same, even exceed to human professionals' abilities. Then, it will cause a lot of human's jobs to be disappeared or some human's jobs will be replaced by owning judgement and mind abilities of (AI) machine men to do.

Hence, future many human's jobs will be replaced by technological labors. Employers choose to buy (AI) machine men to replace human labors. The reasons include (AI) machine men have none unhappy, angry emotin to influence their low efficiencies and low productivities. Their judgement and mind abilities can exceed human's abilities or do any jobs to compare better performance to human's abilities. Consequently, different occupation labors need to prepare to learn how to co-operate with (AI) machine men to let future employers feel (AI) machine men will be human's assistant to assist human to do jobs efficiently when human and (AI) machine men work together. It aims to avoid future employers feel (AI) machine men's judgement and mind abilities can exceed any low knowledgeable and skilful occupation labors, even high knowledge and skilful occupation labors. It means that (AI) machine men are only labors' assistant if (AI) machine mens' judgement and mind abilities are below under to human labors' judgement and mind abilities.

Consequently, to avoid (AI) machine men can replace human to do any simple or complex jobs to cause any future any occupation labors' competitiors. I recommend that it is right time labors ought prepare to learn different skills. So, every individual labor does not only concentrate on one kind of skill. Because supposing one kind of the occupation labor's job duties are replaced by (AI) machine men. If the employee had owned more than one kind of occupation skill. Then, I believe that who can avoid the unemployment threat more easier than the employee only owned one kind of occupation skill, when (AI) machine men had invented to own human's mind and judgement abilities in future one day.

Why does AI machine lecturer can raise
education quality

When (AI) machine men can own human reading and writing and judgement and analytical abilities, then they can replace university lecturers to teach students to raise students' learning abilities absolutely. Then, it bring this question: Why does I machine lecturer can raise education quality? Why do universities prefer to apply (AI) machine lecturer to teach teachers more than human lectuer in university lecturer hall learning environment? Will (AI) university lecturers replace human lecturers to teach students to learn at university lecturing halls popularly? Can (AI) university lecturers replace university human lecturers to teach students more easily and it can let students feel more easily to learn when they are listening what (AI) university lecturers are teaching to them every time university lecture.

In university today, nearly all students need to attend university lecturing hall to listen their lecturer's teaching in every time course. However, many students do not feel interesting to attend university halls to listen human lecturer's teaching. The reasons include, they are busy, so no time to attend lecturer's hall to listen lecturer's teaching; or they feel bore to listen their lecturer's teaching; they feel difficulty to learn; they have confidence to exam and do their assignments, so they feel that they do not need to go to lecturing halls to listen their human lecturer's teaching. However, if one day, (AI) machine lecturers are invented to teach university students to learn and solve their learning difficulties. Can it raise student individual learning interest, due to (AI) machine lecturers' education quality is better than human lecturers' education quality?

What will influence to university students if (AI) machine lecturer can invented to replace human lecture? The influences will include such as below:

First reason: the only way is going to be useful to university lecturers are if all (AI) machine lecturers are well-informed and fully supported to assist human lectuers to teach whose students to let them to listen whose teaching absolutely. So, human lecturers can concentrate on doing any education research and data gathering jobs to prepare for (AI) machine lecturers to help them to explain human lecturers' every time prepared course contents more efficiently. So, (AI) machine lectuers can help human lecturers to share whose teaching time in lecturing halls. Human lecturers' can spend whose hall lecturing time to do whose educational research or other educational gathering jobs absolutely.

The second reason, the human lecturer (Human capital) has ability and efficiency of concentrating on education data gatehering research jobs to prepare to write whose books. When (AI) machine lecturer replace the human lecturer to spend time to attend lecturing hall to teach students. Fo long term, the human lecturer can raise education productivity growth and education quality raising, due to who only concentrate on searching or gathering data to prepare to write whose books to raise their education level.

In macro and micro economic view, the well (AI) machine lecturer educated labor (human capital) is often replaced to human lecturer as one of the critical factors to influence rapid education productivities and educational quality growth to the Asia developing countries' any regions or cities. Because any of these Asia developing countries, such as China, Korea, Philippines etc. countries which need have well educated and knowledgeable lecturer labors to raise any universities' educational productivities and educational qualities growth. So (AI) assistant lecturer factors which ought have close relationship to cause the good or bad future student learning effectiveness and education or learning qualities raising in these any one of Asia developing countries.

The third reason, for the big population of student growth number example, China's student growth rate is larger than school growth rate. If China expect every students have enough chance to study in schools, but university human lectuer numbers are not enough to supply to universities to teach their students. I believe that (AI) machine lecturer is only one kind of teaching method to solve these big population countries' university lectuer number shortage challenge.

In conclusion, in long term, (AI) machine lecturers can solve university human lecturer shortage challenge as well as they can attract many students to attend lecturing halls and human lecturers can raise education quality when they can concentrate on searching or gathering data to prepare their education career, when (AI) machine lectuers replace them to spend time to attend univesity halls to teach students in every university lecturing time.

Future AI machine education market

I believe that when AI (artificial intelligent machine men) which can invented to own to similar to human mind, learning, language, analytical, judgement abilites. Then, which can be applied to any education market service industy. (AI) potential education market service industy includes such as below:

● (AI) university lecture assistant

Future (AI) machine men can assist univerity lectuers to attend university halls to attempt to teach university students for different subjects, e.g. english, math, economic, math, engineering, art, architect etc. different subjects. It depends on the human lectuter who prepares to spend time to teach the (AI) machine lecturer to remember whose teaching subject. For example, the economic lecturer spend one year time to teach the (AI) machine lecturer to learn all economic knowledge. Then, the (AI) machine lecturer can use its machine brain to remember all the human lecturer's economic concepts and prepared teaching economic contents within the one year. Hence, after one year the (AI) machine lecturer can remember all the human lecturer's economic concepts and economic theories and economic contents to prepare to attend university lecturing halls to teach all first year undergraduated first year economic students confidently. It means that the human lecturer's job duties will change to teach (AI) machine lecturer to learn whose economic knowledge to prepare to let the (AI) machine lecturer to replace whom to teach whose university students; so the human lectuer can spend more time to do other research job for whose university education development. Hence, the (AI) machine lecturer can share the human lecturer teaching job as well as the human lectuer can concentrate on spending time to do whose research jobs for whose university education development. This is one both win strategy to university and the lecturer if (AI) machine lecturer is invented to assist future university lecturer's teaching jobs.

● (AI) secondary and primary teacher assistant

In the future (AI) technolgical development, instead of (AI) machine men can be applied to university education aspect. Future (AI) machine men can also be applied to secondary and primary teaching aspect. For example, primary and secondary schools do not need attend classroom to teach students. (AI) machine teachers can replace them to attempt to do teaching job. They only need to spend one year time to prepare to teach (AI) machine teacher to learn how to apply their teaching skill concern their subjects who need to teach to their students, e.g. english language writing and reading and spelling skill, sing song skill, drawing picture skill, calculation skill etc. different studying skill. Then, the (AI) primary or secondary machine teacher can apply the primary or sendary human teacher skills to attempt yo teach whose students. Hence, the primary and secondary human teacher whose duties will change to learn how to teach whose teaching skills to let the (AI) primary or seondary machine lecturer to remember how to apply human skills to teach whose students for different subjects, such as, english writing and reading and spelling language skills, singing songs language skills, math calculation skills etc. Hence, future primary or secondary school teachers who responsibilities will change to learn how to teach (AI) machine teacher teaching skills to prepare to replace them to teach their students in classroom.

● (AI) scientific research assistant

Future (AI) machine men can be applied to science research aspect, instead of school education job. For example, (AI) machine men can be any scientist's assistant, e.g. space scientist, earth or ocean scientist, human or animal behavioral psychological scientist, climate scientist, chemical scientist, drug scientist etc. How can (AI) machine men can be any kind of scientist to assist scientists to do research jobs ? I shall indiate such as below: For space science example, the (AI) machine space scientist can assist human space scientist to gather space data to assist space scientist to research any undiscovered material to cause our earth, even space. Hence, the space scientist only need to teach the (AI) machine scientist to learn how to help them to apply space technological tools to gather data and then enter all data to computer to record, even the (AI) machine scientist can store all space data discovered record to their machine brain every day. Hence, the human space scientist does not need to spend much time to do gathering data job. The (AI) machine space scientist can help whom to do these space data gathering job, then the human space scientist can concentrate on spendin time to do space research job in whose space science laboratory every day.

For earth science example, the (AI) machine earth scientist can help the earth scientist to go to anywhere to gather earth or ocean natural activity data in our earth every day. Then, the earth scientist only need sit in whose earth laboratory to wait the (AI) machine earth scientist to come to whose laboratory to give whose gathering every day earth or ocean natural activity data to do future research job. Hence, the earth scientist does not need to leave whose laboratory to do any data gathing jobs concern earth or ocean natural activities. The (AI) machine earth or ocean scientist had helped him/her to go to our earth or ocean anywhere to do any earth or ocean activities data gathering jobs every day. Hence, the earth or ocean scientist can concentrate on spending whose time to do any research jobs in laboratory. It means that the (AI) machine earch or ocean scientist had replaced whom to do all outdoor original gathering data jobs.

For these human or animal behavioral psychological scientist, climate scientist, chemical or drug scientist, scientist all examples, the (AI) machine human or animal behavioral psychological scientist can help them to do any data gathering job, e.g. the (AI) machine scientist can learn how to help human or animal behavioral psychological scientist to contact human or animal to observe their daily activities and record all their activities data to transfer all these daily activites data to let the human or animal behavioral psychological scientist to do psychological researching analysis only. The

(AI) machine climate scientist can help the human climate scientist to arrive anywhere to observe climate changes and record climate changes daily. Then, the human climate scientist only need to wait the (AI) machine climate scientist's gathering climate change data record from whose machine brain to do climate changing predict research job in climate laboratory every day. The (AI) chemical or drug machine scientist can help the drug or chemical scientist to gather data of new drug or chemical from internet channel every day. So, the human chemical or drug scientist only need to do researching job after the (AI) machine scientist transfers all daily chemical or drug information to let them to know from internet channel. It means that the chemical or drug human scientist does not need to spend much time to gather drug or chemical new data development trend from internet. The (AI) machine chemical or drug scientist had helped them to do data gathering job every day.

Consequently, future (AI) machine men can do education and research aspects of jobs duties and their role are only human scientists or primary or secondary teachers or university lecturers whose assistants either to share scientist's data gathering job or share teachers or lecturers' teaching job.

CHAPTER III

Competitive influences between artificial intelligence and human job raises economic growth

Although, (AI) technology will be popular to applied to different jobs, but it still needs social acceptance to replace some human jobs. Today, it is increasingly common for people to use robots in various situations at home and in retail stores, hotels and hospitals. Robots are classified into several types based on their functionality (service and utility robots or those designed to communicate with humans) and appearance (humanoid robots or mechanical robots). The types of robot to which every country attaches particular important in the advance of robotics, reflects the sense of values and preferences of its population . Thus, (AI) will be applied to replace human to do these above different kinds of job nature. For example, U.S. has the highest level of robot utilization at home and an retail stores with its people being the most enthusiastic about the future use of robots. Otherwise, Germany shows a strong tendency to consider robots for industrial purposes, and its people feel strong to the presence of robots in their households. Japanese accepts to apply" human aid robot" that can communicate with humans and they have a high level of familiarity with robots.

Hence, it implied those three countries have accept (AI) to replace human to do any these kinds of job duty and it will influence these three countries' workers lose their old occupations and who will unemployed absolutely, due to many (AI) robots replace them to do their job duties in the future. Also, US will have many retail service workers or retail warehouse workers are unemployed. Germany will have many manufacturing industry's workers are unemployed. Japanese will have many communication industry workers are unemployed, such as telephone service, shopping center services etc. different kind of service industry's service staffs . It will cause these kind of workers' competitive abilities are lost in themselves countries' jobs that require such skills include software developers, court judges, nurses, high school teachers, dentists and university lecturers, these occupations are still difficult to be replaced by

(AI) robots.

Are robots taking our jobs or making them? In fact, our societies will have unemployment challenges, even (AI) technology has not created before. However, after (AI) robots invention, some of human jobs will be replaced and it can raise many low skillful and low knowledge level worker unemployment number. However, I think that high productivity driven by increasingly powerful IT -enabled machines is the causes of global labor market problems and accelerating technological change will only make those problems worse.

IT technology brings this question: Are robots killing human's jobs or benefiting human's jobs? I suppose that there is a limited amount of labor to be done. The implication is that technology can create unemployment by displacing workers, such as (AI) invention, because the more efficiently worker work (using machines or (AI) robots), the loss work there is for workers to do. Even, any new jobs will be better done by machines or (AI) robots, and unemployment will still skyrocket. How do we know that humans will always be better at some work, or more importantly, enough work, than machines or (AI) robots, e.g. human drivers drive more safe or careful to compare (AI) robot drivers. But, the challenge is that it is not ensure that (AI) robots drivers must not drive careless to cause the chance of accident occurrences more than human drivers. However, technological change can be beneficial to innovation, automation and increasing productivity for businesses.

Consequently , it may seem machines can hurt wages and job for low skillful, less educated workers. Also, high educated workers are likely as less educated workers to find themselves displaced and devalued, and more education may create as many problems as it solves. Thus, in negative influence, automation effects on particular jobs shift workers to other jobs that are equally or more desirable. Workers may be highly compensated for possessing human capital that is specialized to a labor market. If technological advance is very rapid, such as (AI) invention, causing a large and very rapid drop in demand in a large labor market, the economy may not be able to absorb the sudden surplus of labor in a short period of timer when (AI) robots are popular to replace some workers to do some occupations in global societies.

For example, self-driving vehicles threaten to send truck drivers to the unemployment office. Computer programs can now write journalistic accounts of sporting events and stock price movement. There are even computers that can grade essay revolutionize some part of teaching jobs. Hence, (AI) robots will have possible to replace human brain to do any judgement, argument, and mind job duties. It implies some occupations which need human' mind will be threaten by (AI) robots, e.g. author, accountant, nurse, engineer. Thus, (AI) robots will have possible to replace some professional and high educated workers' jobs in the future.

But, technology can create new nature of jobs in possible. For example, a 60 minutes program indicated technology is putting new categories of jobs in the sites (sic) of automation, the 60% of the workforce that makes its living gathering and analyzing information. Also, recession: technology kills middle -class jobs that overall technology is eliminating for more jobs than it is creating by (AI) technology. Hence, human's brain work may be assisted by 60% of (AI) gathering and analyzing information for some occupation , e.g. space scientists, ocean scientists, earth scientists etc.

However, I believe the (AI) invention and human job competition may influence global productivity change. Productivity is economic output per unit of input, the unit of on input can be labor hours(labor productivity), but if (AI) robots replace human job, then the unit of input may be (AI) machine hours (AI) robot productivity or all production factors including labors, machines and energy (total factor of productivity). Producing more output with less input can take several forms.

The traditional notion of productivity is a form reorganizing production and/or using better or more technology to produce more output per worker hour. But when (AI) robots invention, the form can be reorganizing production and/or using better or more (AI) robots to produce more output per (AI) robot hour. Hence, if the firm apply (AI) robots to produce its products. Then , productivity improvements in the firm may result in less workers employment, due to (AI) robots replace more worker number to achieve more productivity improvement, it has economic benefits (less factor of production) , but more production in long term.

Thus, (AI) robots can help any firm to achieve productivity improvement in long term, for example, if unproductve farmers move to the city and start working for high-tech. manufacturers. The shift effect can be more dynamic and disruptive as low-productivity industries lose out in the marketplace to high -productivity industries and the compositional mix of the economy changes. Thus, in the long term (AI) robots can also be beneficial to high productivity industries to bring the mix of economy positive changes.

Moreover, automation will also produce some new jobs in firms that sell the new robot or other labor-saving technology. This means that, in general, there will be shift in the economy in the direction of higher-skill and higher wage jobs. Even if the (AI) robot invention country, US becomes a leader in (AI) robots producing productivity-enhancing technology, it will experience a growth in jobs serving foreign (AI) robots product buyers. Hence, (AI) robots can also create (AI) salespeople, (AI) manufacturing workers , (AI) inventors, scientists, (AI) software designer etc. occupations, when if all society does is move workers from insurance firms, restaurants and car factories to robot factories, productivity will have remained the same to create job needs for insurance, restaurant and car manufacturing worker service occupations for (AI) software designer, (AI) service robots manufacturer, (AI) service robot seller etc. related (AI) service robot product occupation created in (AI) robot technology job market. Hence, (AI) invention also create new (AI) technology job chance. (AI) impacts management job market.

In future, organization management will be changed from (AI) introduction. Division of labor will change and collaboration among humans and machines will increase. Companies will have to adapt their training, performance and talent acquisition strategies to account for a new found emphasis on work that hinges on human
judgement and skills, including experimentation and colloboration.

How (AI) impacts any organizational administrative management work? (AI) 's greatest impact will be on administrative coordination and control tasks, such as scheduling, resource allocation and reporting, (AI)-driven will place a higher premium on what we call " judgement work", the application of human experience and expertise to critical business decisions and practices when information available is insufficient to suggest a successful course of action. This kind of work will require new skills and mindsets; replacing people with machines is not goal in itself. When, artificial intelligence enables cost-cutting automation of routine work, it also empowers value -adding augmentation of human capabilities.

Thus, administrative and routine tasks, such as scheduling, allocation of resources, and reporting, will within intelligent machines, responsibilities that have long been reserved for humans. For instance, a typical store manager or a lead nurse at a nursing home must constantly juggle shift schedules, accounting for staff members' absense owing to illness, vaction, time or sudden departures. Many of these tasks will be automated by (AI).

Imagine (AI) writing management monthly reports, it is not a distant dream. Leading news providers and Wall street banks are now using (AI) report generators to write news and analytical reports by drawing on quantitative data. The associated press, for example, expanded its quarterly earnings reporting from approximately 300 companies to nearly 3,000 with the help of (AI) powered software robots, freeing up journalists to conduct more investigative and interpretive reporting. For another example, Jobalime, a job-placement site, uses intelligent voile analysis algorithms to evaluate job applicants. The algorithm assesses paralinguistic elements of speech, such as tone and inflection, products which emotions a specific voice will elicit, and identifies the type of work at which an applicant will likely excel. In the future , (AI) machines can be applied to assist some kind of office administrative jobs duties. It's attractive to office managers to achieve more accurate judgment to do any administrative matters when who can be assisted from (AI) machines. Thus, managers need to spend time to learn how to apply (AI) machine to assist them to do more accurate judgement, and better informed choices. (AI) robots can be applied to improve the speed quality and cost of available products and services, instead of applying on productivity improvement and administrative improvement aspects. Thus, they may also displace large numbers of workers. This, possibility challenges the traditional benefits model of trying health care and retirement savings to jobs.

In an economy that employs dramatically fewer workers to deliver benefits to displaced workers. For example, the worldwide number of industrial robots has increased rapidly over the past few years. The fall prices of robots, which can operate all day without interruption, make them cost- competitive with human workers. In special consideration, in the service sector, computer algorithms can execute stock trades in a fraction of a second, much faster than any human. As those technologies become cheaper, more capable, and more widespread, they will find even more applicants in an economy.

Consequently, (AI) technology brings unemployed number increasing many businesses continued automating their operations rather than hiring additional workers. A trend among technology companies that receive massive valuations with relatively few workers. For example, in 2014 year Google was valued at $370 billion with only 55,000 employees, a tenth the size of AT & T's workforce in the 1960 year. Hence, if automation technologies like robots and artificial intelligence make jobs less secure in the future, there needs to be a way to deliver benefits outside of employment " flexi security" or flexible security is one idea for providing healthcare, education and housing assistance whether or not someone is formally employed.

In conclusion, (AI) and robots technology will raise unemployment to some occupations when (AI) replaces same industries' workers job duties in our societies in the future, but it also create new jobs to raise employment in any related (AI) robots and automated machine products in (AI) manufacturing. (AI) design, (AI) sale self-related industry, when (AI) replaces same industries' workers' job duties.

(AI) journalism, media publishing, digital
communication technology trend

How to apply (AI) technology in digital communication journalism media, publishing industry? Some scientists indicate future (AI) and digital technology may consist such as: voice driven assistants, emerge. For example, Amazon e book publish applying digital technology and (AI) auto printing technology to sell e books to let readers to listen any e book content by (AI) voice driven speaker when they turn on computer to read e book contents; capable phones start to unlock the possibilities of 3D image of mobile story telling. New smart wearables include ear buds that handle instant translation and glasses that talk and hear. China and India will become a key focus for digital growth with innovations around payment online identity, and artificial intelligence. Thus, future (AI) technology can be applied to 3D image mobile story telling, online payment method to dealt online transaction publishing industry.

Thus, future (AI) technology can be applied to online e book publishing industry to make sound books to let readers feel more attractive . Such as Amazon publish has published sound e books to attract readers to choose to read any its books from online. Also, (AI) technology can also be applied to communication industry. For example, some online pure-play news, opinion and entertainment websites. It is a digital communication media, e.g. online journalism blog (AI) technology can be applied to visual storytellers to let online book readers to enjoy to listen to watch and send any online electronic book contents more attractive. Thus, future (AI) technology will be popular to assist any electronic book publishers to publish visual and sound talking storybook to let readers who can watch motive image and listen and read words from e books more attractive.

Thus, (AI) technology can be applied to internet ecommerce publishing or media industry to help any electronic book publishers to publish sound, image motion electronic book to attract global readers to read, even (AI) technology can be applied to digital entertainment industry, e.g. electronic 3D image virtual video games, computer games. It can be also applied to education industry, e.g. the first true digital native generation and are the native speakers of the digital language of computers to let student to learn different languages or translate words to compare to classroom learning more easily. It can be also applied to communication industry, e.g. (AI) mobile phone. Hence, it seems (AI) technology can be applied to publishing, communication, education , entertainment etc. different industries in the future. (AI) technology will be one kind of tool to satisfy human's daily life needs in the future and these industries has one characteristics is that they need to apply internet to operate to operate to do online business.

Thus, it has three trends of (AI) technology and internet technology need to be linked to achieve one kind of attractive technological business to satisfy client's needs. These three trends as below: All consumer trends involve the internet. It will be many consumer's online habits, shopping, working, socializing, watching TV, studying, travelling, listening. Thus, (AI) music, eating and exercising are just a few examples. This is happening because human usually use mobile broadband or Wi-Fi, rather than cables. Thus, (AI) technology will be applied to mobile to satisfy client's need absolutely.

The mobile phone can be more popular to be used more than computer or laptop tools. The reasons are because women dive the smartphone market by defining mass-market use. But as the speed of technology adoption increases mass market use becomes much quicker then before. Successful new technological products and services , such a (AI) mobile phone products now reach the mass market in popular use. It means that the time period when early adopters influence others is shorter than before. Also, since new products and services increasingly use the internet mass markets are not only faster , but are also more important than ever to consumer themselves. Most internet services become more valuable to individuals when many use them. Thus, it causes why (AI) mobile phone will be popular to be used.

Since, new products and services increasingly use the internet, in the future several trends focus on (AI) smart phone users. Consumers' familiarity with using smartphone apps. Essentially, the technologies will bring other related (AI) and internet service needs, e.g. sound and image emotion e book needs, (AI) mobile communication needs, e-virtual games or e-3D image virtual games etc. entertainment activities needs with such a large part of the world's population now online, it is clear that there is strength in numbers.

Thus, (AI) imagines , if future any (AI) and internet related services or products new technology is easy to use and inexpensive, when the latest products reach the mass market almost as quickly as they reach the early adopters and industry experts. I believe that any (AI) and internet related products or services must be popular to accept to consume for entertainment or useful aim. For example, with major players including Apply, Facebook and Google had invested (AI) technology to develop their businesses. (AI) technology has the potential to disrupt everything in the coming years, from the lives of connected consumers to every industry (AI) will be an alternative route for brands to reach consumers with convincing and relevant messages. Digital technology will assist of the future, then it can improve technology to bring this effect, such as sophisticated software machine learning and speech recognition effective. Hence, Google, Facebook , Yahoo web site service companies can apply (AI) technology to help other companies to advertise their businesses, such as travel, retail, and education etc. industries more attractive. (AI) technology can be applied to internet company to be aware and familiar enough to drive among mainstream consumers, it can create online experience to travel, retail , education and other entertainment needs to online consumers to seek their entertainment needs more easily. Hence, in the future (AI) technology and internet related entertainment service needs will be raised in this (AI) and online consumption market.

● (AI) healthcare service industry development

In the future, (AI) medical internet technology tool can be applied to assist individual's health at the center of their focus, e.g. smartwatch compatible mobile app. patients can let personalized reminders for taking their medication snap pictures of their prescriptions to expedite refills, and scan their insurance card. So that, store clerks are prepared with up-to-date patients' information . (AI) owned health operated technological clinics can help patients to receive treatment for minor illnesses, flu shots, cholesterol screenings and more than a dozen other medical services, all of which can be patients who can't make it to a physical location. (AI) healthcare services organizations can provide various telemedicine services. So, patients can receive care via phone or video chat.

For example, one London-based intelligent Brewing company has developed an (AI) system to continuously collect and incorporate customer feedback, which the system itself uses to brew ne various of the company's beers. Thus, the beer clients can give feedback to talk to the algorithm (AI) machine, whenever or anywhere who're drinking the beer. It is such any healthcare services organizations can apply (AI) machine to collect patient's feedback to talk to the algorithum (AI) machine whenever or anywhere who're eating any medicines. So doctors can know every patient's health conditions any time. If the patients feel uncomfortable, the doctor can know from (AI) machine notification to decide whether the patient needs to eat another new medicine or keep to eat same medicine is better. Hence, (AI) medial internet technological body check report machine will be proper to be needed to serve any hospitals' patients in the future.

However , it brings this question. How can (AI) medical internet technological body check report machine apply to hospital more efficient? The essential new medicine co-workers for the health service digital age health service leaders need apply (AI) medical report machines and artificial intelligence to the newest recruits to the workforce bringing new skills to help health service staffs do new jobs and reinventing what's possible, building the health service workforce for today's digital health service demands for patients. Thus, technology-driven health service model innovation from the health service organization outside in and providing digital health service ecosystems for patients to use the (AI) health service equipment will be popular to be accepted to be used.

● I Robot and internet things future machine
men invention

Nowadays, there are some company, which apply internet and (AI) I Robot technology to do any similar human job nature. For fishing industry example, one company, known for creating the Roomba, I Robot is now working with marine conservationists to launch an ocean-patrolling intelligent robot to hunt and manage invasive species, protecting native populations. And evolved industries like precision agriculture are ramping of our increasing population. Area of practice that once seemed impossible to digitize are fundamentally changing because of the impacts of (AI), internet of things capabilities and big data analytics, which have many potentially positive impactions for society.

For textile industry example, automation is nothing new, it has shaped the workplace to replace human jobs to boost productivity in the textile industry. Textile machines have had a generally positive impact over gears, creating value and allowing textile workers to take up more rewarding age will likely continue to create opportunities and lead to new textile industries, companies and textile occupations. It may also compensate for a demographically driven slowdown in the growth of the textile workforce. The future impact of textile (AI) and automatic and internet link is somewhat uncertain. It seems textile industry will be trend to accept (AI) textile workers and internet of thing to replace traditional manual textile workers to produce any shirts, cloths etc. wearing products in factories popularly, during the (AI) textile machine and internet thing technology can be invented to reach the mature stage in the future.

For factory worker transportation job example, they have also expanded their influence, migrating from the factory floor to the service sector and taking the place of humans in a range of activities from financial transactions to transport route optimization. Further (AI) machines and robots are increasingly programmed to learn, meaning they improve with time and undertake cognitive activities. Hence, (AI) machines and internet technology enable automation of work activities to raise factory workers' efficient and performances, also factories can reduce manual worker numbers, due to (AI) machine workers' assistance.

Future, (AI) robotics technologies and internet technique have these different kinds of characteristics: For soft robotics example, it is non-rigid robots construct with soft and deformable materials that can manipulate items of varying size, shape and weight with a single device. For swarm robotics, it coordinated multi-robot systems often involving large numbers of mostly physical robots. For touch/factile robotic example, it robotic body pails (often biologically inspired hands) with capability to sense, touch , dexterity robots example, serpentine robots with many internal degrees of freedom to threat through tightly packed spaces for humanoid robots example, robots physical is similar to human being often bi-pedal that investigate variety capable of performing human tasks , including movement across terrains, object recognition, speech sensing etc. For autonomous cars and trucks example, it is capable of operating with a human pilot, e.g. the unarmed general atomics Predator XPUAV with roughly half the wingspan of a Boeing 737 can fly autonomously for up to 35 hours from take-off to landing, for unmanned aerial vehicles example, flying vehicles capable of operating without a human pilot, the unarmed general atomics predator -XPUAV , with roughly half the wingspan of a Boing 737, and fly autonomously for up to 35 hours from take off to landing, for (AI) chat bots example, (AI) systems designed to simulate conversation with human users, particularly those integrated into massaging apps.

In Dec. 2015. the general service administration of the US Govt. described how it used a chat bot named Mrs. Landingham (a character from the television show the west wing) to help onboard new employees. Finally, for robotic process automation example, class of software robots that replicates the actions of a human being interacting with the user interfaces of both software systems. Enables the automation of many back-office work flows without requiring expensive IT integration . Hence, future (AI) robot machine men will have different functions to be applied to different industries to use in possible.

Statistics Denmark shows that (AI) automation potential robots will influence few jobs are completely automatable , but close to half consists of 40% automatable tasks: It showed example occupations include share of automated, such as brewing machine operators are more than 80%, logging equipment operators are more than 50%, roofers , stock tasks clerks, travel agent are more than 50%, farmers , nursing assistants are more than 30%, physicians, teachers , managers are more than 10%.

For example, humans perform a wide variety of tasks from planting corn to examine spreadsheets, meeting clients and lifting crates in a store. Each of these actions requires a combination of innate or acquired capabilities, internet technique assistance, ranging from social perceptiveness to fine motor skills and natural language understanding. To understand and map automation feasibility by existing technology. Mc Kinsey has developed a framework of 18 technical capabilities that can substitute tasks performed by humans. The capabilities are grouped in five categories: sensory, cognitive, language, social and emotional and physical. So, it seems (AI) robot machine men and internet technique will have possible combination to invent to own human's emotion , language, learning, task skill abilities.

Mckinsey global institute analysis also showed current technologies have achieved different levels of human performance across 18 capabilities include: sensory perception, autonomously infer and integrate complex input using sensors, cognitive capabilities reorganizing known patterns/ categories supervised learnings, generating novel, logical reasoning/ problem solving, optimization and planning, creative, information retrieval, coordination with multiple agents, output articulation/presentation, national language processing, social and emotional capabilities-natural language understanding, social and emotion sense, reasoning output, physical capabilities-fine motor skills, navigation mobility. Hence, it seems (AI) robots and internet technological will combine to invent to own human' some skills to replace human to do some kind of tasks in possible.

In conclusion, future (AI) robot and internet will be needed to link to cooperate together to raise human's work efficiency in popular.

How artificial intelligence replaces human job possibility

What is the risk of automation for jobs to replace human job? In recent years, there has been a revival of concerns that automation and digitalization night after all result in jobless future. As I argue, this might lead to an overestimation of job (AI) automate , as occupations labelled as high-risk occupations often still contain a substantial share of tasks that are hard to automate.

For example, when the share of (AI) automatable jobs is 6% in Korea, the corresponding share is 12% in Australia. Differences between countries may reflect general differences in workplace organization, differences in previous investments into (AI) automation technologies as well as differences in the education of workers across countries. I also discover that (AI) automation and digitalization are unlikely to destroy large numbers of jobs. But, however, low qualified labors are likely to raise costs as the (AI) automate of their jobs is higher compared to highly qualified workers.

In fact, (AI) technology will influence some new technology to replace some human's job, such as driverless car, the largely autonomous smart factory , service robots or 3D printing. These technologies are driven by advances in computing power, robotics and artificial intelligence and ultimately redefine what type of human capabilities machines are able to do.

Hence, question brings whether (AI) invention will influence general human jobs to be replaced by (AI) autonomous jobs? Whether will the potential foe automation with actual employment loss? In particular, the technical possibility to use (AI) machines rather tasks need not mean that the substitution of humans by machines actually takes place.

Whether (AI) technology replaces human's some job, it is beneficial to our society or not. Instead, machines are increasingly capable of performing non-routine cognitive tasks, such as driving or legal writing . In particular, advances in the field of machine learning (ML), e.g. computational statistics and visions, data mining, artificial intelligences allow for automating cognitive task, when the use of (ML) in mobile robotics (MR) also allows for automating certain manual tasks. So, it seems, (AI) technology can replace some labor job, e.g. warehouse transportation, even mind's job, e.g. legal writing, driving in possible.

For example, if (AI) automatic non-manual driving can reduce hurt or death risk, it is beneficial to our society, or (AI) automatic robots can more any heavy things (products) in warehouse safely. Then, it can reduce the warehouse labor's bodies hour risk, it is beneficial to the workers. Even, if (AI) robot can write any legal documents, no any word errors in short time. It is beneficial to the law companies , but it also bring unemployment chance, due to these jobs can be replaced by (AI) robots to do in the future. Hence, it will cause some occupation to be disappeared, due to (AI) robots can do our these kinds of jobs in the future.

Frey & Osborne (2013) reported these kinds of occupations will be replaced by (AI) robots in possible. They include computer, engineering, financial, management, legal , art and medium, community service, education, healthcare practitioners and technical service, sales and related, office and administrative support, farming, fishing and forestry, construction and extraction, installation, maintenance, and repair , production, transportation and material moving. It seems our future some professional occupations will have possible to the replaced by (AI) robots to replace, instead of labor jobs. Hence, (AI) robots technology will have much trend to replace high knowledge or low knowledge skillful labors in the future.

In conclusion, it implies that only using information on task-usage at the individual level leads to significantly lower estimates of jobs " at risk", some workers in occupations with according to high automate nevertheless often perform tasks with are hard to automate. Why can (AI) replace human to do some kinds of jobs? (AI) artificial intelligence refers to the ability of a computer or a computer enable robotic system to process information and produce outcomes in a manner similar to the thought process of humans in learning, decision making and solving problem. By extension, the goal of (AI) systems is to develop systems to capable of tasking complex problems in ways similar to human's logic and reasons who feel in our future. Hence, it means future (AI) robots has effort to replace human to do any jobs in possible.

(AI) directions for future non-manual
control road vehicles market

Future road vehicle products and technologies must meet social, economic and environmental protection and driving safety goals , and satisfying market requirements for mobility, accident reducing, performance, cost desirability. Thus, (AI) auto-non manual control vehicles need to be followed this direction to invent. To satisfy future driver's safety of needs, enhanced vehicle speed desired functional performance of road transportation system, required and desired technological response, including research needs. It is long term up to 20 years vision, for (AI) auto non-manual research. Thus, (AI) auto non manual vehicle manufacturers need often to revise their (AI) vehicles functions to raise to improve their system performance and driving industry driver's needs, e.g. private drivers need or public transportation driver's need or business client's need. Hence, future (AI) transportation will need have individual driving consumer and business driving consumer both targets.

Thus, future (AI) automation manual control vehicles need to deliver high impact technology solutions to meet social , economic and environmental and safe goals. Engine needs to be improved efficiency , performance, drivability, reliability, durability and speed-to-market together with reduced emissions and cost; hybrid, electric and alternatively fuel (AI) non manual control vehicle technology development, leading to new fuel and power systems, such as hydrogen, fuel cells and batteries, which satisfy future social, economic and environmental and safe goals. Software, sensors, electronics and telematics technology development are needed to be lead to improve vehicle performance, control and adaptability, intelligent , mobility and security, structure and materials technology development, leading to improved safety, performance and leading to flexibility with reduced cost and environmental pollution to achieve the (AI) non manual drivers to feel (AI) vehicle performance, auto control and adaptability is better to compare traditional manual driving vehicles.

In fact, in traditional manual driving market, Japan and USA had had over 80% of world car production by six major global groups. In the future, it is possible only USA can dominate (AI) non manual auto driving vehicle manufacturing market if Japan had no effort to manufacture any (AI) auto non manual control vehicles. So, it means that it is only Japan is USA potential (AI) auto non manual control vehicle manufacturing competitors. Also, it means that it is only USA has effort to export (AI) auto non manual control vehicles to global (AI) auto non manual control vehicle market.

Thus, in long term, (AI) non manual control vehicle product market development, USA (AI) vehicle manufacturers will have these requirement to win new technological competition to traditional manual control vehicle. The requirements include: low cost fuel, low carbon, fuel cell and telematics technologies, the technological roadmap function, such as detailed consideration clear provision of other important areas to the drivers. When the (AI) non manual auto drivers are sitting in the non manual control auto vehicle. Although, who does not need to drive, but who need to know how to go to anywhere by electric road map show clearly. So, the driver won't lose direction and he/she can know the (AI) non manual control vehicles is driving to anywhere in any time, even when who is sleeping.

In the future, the (AI) non manual control vehicles need to be invented to satisfy any business , transportation clients' needs, instead of individual clients needs, e.g. cans, trucks, buses, emergency and utility vehicles, trains, trams etc. Hence, technological road mapping is one important tool to help any business, transportation (AI) non manual control vehicle clients. Technological roadmap is a technique that is used in industry to support strategic planning for (AI) non manual driving vehicles in the future. Electronic road maps generally take the form of multi-layered time based charts, linking technology developments to future (AI) non manual control vehicle market requirements.

Technology road mapping is a flexible technique and the roadmap architecture and process for developing the roadmap most generally be customized to meeting the particular aims. Why technology roadmap will be popular to (AI_ non manual control vehicles. It's advantages include: It is a technology solutions and options that can enable the performance targets to be achieved engine hybrid, electric and alternatively fueled vehicles, software, sensors electronics and telematics, structures and materials design and manufacturing process. It is road transport system performance measures and targets tool, in response to the trends and get (AI) non manual control vehicle drivers to get society, economy, environment protection, low cost driving benefits, also it can help any transportation clients to know how to go to anywhere clearly. Hence, technological road map will be one good tool to assist (AI) non manual control auto vehicle to develop future road driving market.

Reference(source)

Frey & Osborne (2013), The future of employment: How susceptible are jobs to computerization? University of Oxford.

Mckinsey Global Institute Analysis

Statistics Denmark, Global automation impact model, Makinsey analysis

(AI) development second stage

(AI) -driven automation industry development

(AI) -driven automation industry will create wealth and expand economy growth to any countries, but it will be accompanied by changed in the skills that workers need to learn. One of main ways that technology increases productivity is by decreasing the number of labor hours needed to create a unit of output. It implies (AI) technology will influence low educated and low skillful labor number to be decreased (reduction employment number).

In contrast, technological change tended to work in a different direction throughout the nowadays. The advance of computer and the internet raised the relative productivity of higher skilled workers. So, routine-intensive occupations that focused on predictable tasks disappearance, such as switch board, operators, filming checkers, travel agents and assembling line workers etc. were particularly replaced by new technologies.

However, today, it may be challenging to predict exactly which jobs will be most immediately affected by (AI) driven-automation. The reason is because (AI) is not a single technology, but rather a collection of technologies that are felt unevenly through the economy to influence job changing both negatively and positively. In positively view point, (AI) driven-automation will make many workers more productive and increase demand for certain skills. Consequently, new jobs are likely to be directly create in areas , such as the development and supervision of (AI) as well as indirectly created in a range of areas throughout the economy as higher incomes lead to expanded demand. Otherwise, in negatively view point, many traditional human needed (demand) skillful jobs will be threatened by automation are highly concentrated among lower-paid, lower-skilled and less -educated workers. It means automation will cause pressure on demand for this group, pressure and employment, if (AI) can replace the low skilled and less educated workers' jobs. Thus, (AI) will have negative influence to impact on the labor market.

(AI) capabilities will enable automation of some tasks that have long required human labor. Why can (AI) replace some simple human jobs? For example, advances in robotics are expanding machines' abilities to interact with and sharp the physical world. Combined , (AI) and robotics will give rise to smarter machines that can perform more sophisticated functions than ever before and brings more advantages that humans have exercised. This will permit automation of many tasks now performed by human workers and could change the shape of the labor market and human activity.

How (AI) influences labor market

Today, it may be challenging to predict exactly which jobs will be most immediately affected by (AI)-driven automation. Because (AI) is not a single technology, but rather a collection of technologies that are applied to specific tasks.

Some specific predictions are possible based on the current (AI) technology. For example, driving jobs and house cleaning jobs, bank counter service jobs, telephone enquiry service operators. Restaurant cooking jobs, simple accounting record service jobs etc. that require relatively less education to perform. Advancements in computer vision and related technologics have made the feasibility of fully appear more likely, potentially displacing some workers in driving-dominant professions. Seemingly similar robot, for which the operational tasks is less specific of navigating to a specific destination when following a set of given rules and preserving safety.

In the future, the effects of (AI) on the labor market in the decade ahead will continue the trend toward skill-biased change that computerization and communication innovations have driven in recent decades. Thus, some human driving occupation will be disappeared or replaced by (AI) automation driven. For example, bus drivers, light truck or delivery services drivers, heavy and tractor-trailer truck drivers, school drivers, tax drivers, travel bus drivers.

However, (AI) technology could enable some workers to focus time on other job responsibilities, boosting their productivity, and actually raised wage growth among those still holding the reshaped jobs. For example, salespeople, who currently spend a considerable amount of time driving could find themselves able to do other work when a car drives them from place to place, or inspectors and appraisers could fill out paperwork, when their car drives itself. This (AI) -driven technology should make these workers more productive, with (AI) -driven technology serving as a complement, not a substitute. New jobs will also likely be created, both in existing occupations cheaper transportation costs with lower prices and increase demand for products and all the related occupations, such as service and fulfillment, and in new occupations not currently foreseeable.

What kind of jobs will be created by (AI) technology? Predicting future job growth is extremely difficult, due to it depends on technologies or substitute for existing today as well as they may complement or substitute for existing human skills and jobs. However, (AI) will also lead to substantial indirect job creation to the degree it raises productivity and wages, it may also lead to higher consumption that would support additional jobs from high-end draft production to restaurant and retail. The future(AI) " augmented intelligence", the technology's role is as assisting and expanding the productivity of individuals rather than replacing human work. Thus, based on the biased-technical change framework, demand for labor will likely increase the most in the areas where humans complement (AI) automation technologies. For example, (AI) technology , such as IBM's Watson may improve early detection of some cancers or other illnesses, but a human healthcare professional is needed to work with patients to understand and translate patients' symptoms, inform patients of treatment options, and guide patients through treatment plans. Shipping companies may also partner workers who pick up and deliver products over the last feet with (AI) enabled autonomous vehicles that move workers efficiently from site to site. In such cases, (AI) augments what a human is able to do and allows individuals to either be move effective in their specially task or to operate on a larger scale. Thus, it seems (AI) technology will also create new jobs, raise productivities and workers' efficiencies.

Redefining management in the workforce of artificial intelligence
In the future, due to artificial intelligence influences to some kind of human jobs nature. So, the kind of human jobs of management methods will also need to change to adapt the artificial intelligence technology input to their organizations. It will cause challenges for every executive and manager if who won't have effort to manage their teams how to apply artificial

intelligence technology to work efficiently and easily. For example, division of labor will change among humans and machines will increase. Thus, companies will have to adapt their training performance and talent strategies how to emphasize on work that how to make human judgment and skills and experimentation. Thus, (IA)'s greatest impact will be on administrative coordination and control tasks, such as scheduling , resource allocation.

In fact, mangers will encounter this challenges: How to apply human experience and expertise to judge critical business decisions and practices when the information available is insufficient to suggest a successful course of action? Due to this kind of work will require new skills and mindsets. I shall indicate these change management methods to adapt (AI) technology. Such as: administration and routine tasks, scheduling , allocation of resources and reporting will fall within the intelligence machines, responsibilities that have long been reserved for humans. For example, a typical store manager or a lead nurse at a nursing home most constantly arrange shift schedules, accounting for staff members' absences owing to illness, vacation time or sudden departures.

Thus, the managers need to learn how to arrange new division of labor within the organizations after (AI) technology had been implemented to the organization. Artificial intelligence is currently influencing into once considered exclusive to humans: assessing and acting on human emotions and personality traits. The influences to managers need to change their strategies to adapt (AI) technology implements include such as below:

Firstly, managers need to spend the bulk of their time on coordination and control tasks from intelligent system implements. Their time spending on these major three aspects from impact of intelligent system: coordinate and control, solve problems and collaborate and people and community , strategy and innovation three aspects. Thus (AI) will influence managers need to change their judgment method to teach whose teams how to adapt the (AI) system operations in any organizations.

Secondly, (AI) will influence top, middle and low level management needs to change to adapt the (AI) technology operations to any owned (AI) technology organizations in the future. Intelligent machines must be trained in context. Just like humans , on-the-job training is a requirement for such machines because they typically arrive with only very general capabilities. To get the most from (AI), managers at all levels must participate in the instructional experience and in the learning process and provides managers' familiarity with such systems on these aspects, e.g. How the system works and generate advice, how the system has a proven track record , how the system provides convincing explanations , how the system can make simple rule- based decisions.

Thirdly, managers need to learn how to make judgment more accurate (AI) systems assistance. Although (AI) will invariably take on more routine work and even augment human decision-making, it won't judgment work, the application of human experience and expertise to critical business decisions when the information available is insufficient to suggest a successful course of action or reliable enough to suggest an obvious course of action. For a sense of the nature of judgment work, consider big data marketing and sales analytics. Such analytics often provide insights that can inform promotional campaigns, including predicting which promotions will generate desired sales brand further into the future, marketing executives need use judgment, combining analytics with their own and others' insight and experience.

The application of experience and expertise to critical business decisions and practice represents the real value of human judgment. But, when artificial intelligent machines are implemented to any organizations to assist the low, middle and top level management to make any business judgment. These forms of judgment work that managers can gather data interpretation, idea development more absolute from (AI) machine assistance. Thus, why these level management executives need to learn how to apply (AI) machines to help them to make any business judgment more accurate.

How (AI) influences organizational change

Consequently creative and social intelligence will be in even greater demand as (AI) makes in management and the workforce. This development will represent a long term trend in labor markets , one characterized by intensifying demand and reward for social skills with a growing desire for creative capabilities, managers will seek to fashion of ideas and hypotheses from inside and outside of the enterprise to shape solutions to their most pressing business problems. Thus, (AI) will influence overall organizational team members who have chance to participate any decision to make more accurate business judgment.

Many managers mistakenly view judgment work as only an individual discipline, failing to appreciate that it can also involve decide interpersonal and organizational practices. In more complex settings, judgment is typically a collective outcome of individuals' and teams' diverse perspectives, insights and experiences. And often , the resulting choices are better informed than decisions that an individual would have arrived at on his or her own.

Thus, when any organizations apply (AI) technology to assist managers to gather data and ideas to make any judgment. In these cases, organizations can create the conditions for effective collective judgment by establishing structures , such as " shadow advisory boards" that prompt managers and employees to source and synthesize multiple perspectives. Thus, a traditional organization (firm) might freshen its thinking is t put together a shadow advisory board, comprised of young, digital people who can apply (AI) machine assistance to make judgment work more accurate whether related to people development, problem-solving or strategizing and innovating for considerable degrees of creative and social intelligence.

Thus, on the one hand, (AI) technology machine augmentation and automation can give these advantages to human (organization managers) , e.g. developing people and community, solving problems and collaborating, coordinating and controlling work, shaping strategy and leading innovation. Besides, on the other hand, the next generation managers need have these individual attitude to treat intelligent machines to be as colleagues.

When, judgment is a human skill, intelligent machines can accelerate human learning that supports it, assisting in data -driven simulations, scenarios and search and discovery activities. Focuses on judgment work, some decisions require insight beyond what data can tell them. This is the sweet sport for human judgment, the application of experience and expertise to critical business decisions and practices. Thus, managers will also need to find ways to learn how to use digital (AI) technologies to tap into the knowledge and judgment of partners, customer external stakeholders and role models in other industries after the (AI) machine had been implemented to the organization.

Future works change:
Automation, employment
and productivity

Human future " micro to macro" industry trends will be affected business strategy and public policy by (AI) technology. In the future (AI) technology will influence those six themes: productivity and growth, natural resources, labor markets, the evolution of global financial markets, the economic impact of technology and innovation and urbanization. However, (AI) technology will bring economic benefits of tackling gender inequality, a new global competition, Chinese innovation and digital globalization.

Nowadays, advances in robotics artificial intelligence, and machine learning are in a new age of automation, as machines match or outperform human performance in a development to any countries. For example, automation of activities can enable businesses to improve performance by reducing errors and improving quality and speed, and in some cases achieving outcomes that go beyond human capabilities. For example, some research indicated automation could raise productivity growth globally by 0.8 to 1.4 % annually; more than 2,000 work activities across 800 occupations. When less than 5% of all occupations can be automated using demonstrated technologies about 60% of all occupations have at least 30% of constituent activities that could be automated. Many occupations will change that will be automated away: Activities most susceptible to automation involve physical activities, in highly structured and predictable environments, as well as the collection and processing of data. They are most prevalent in manufacturing , accommodation and food service and retail trade and include some middle-skill jobs. For example, such as natural language processing is a key factor. Beyond technical feasibility, the cost of technology competition with labor including skills and supply and demand dynamics, performance benefits including and beyond labor cost savings, and social and regulatory acceptance will be affected by (AI) automation technology. Thus, (AI) automation will impact to influence global employment in those aspects as below:

Firstly, assuming that people are displaced by automation will find other employment. The anticipated shift in the activities in the labor force is of a similar order as the long-term shift away from agriculture and decreases in manufacturing share of employment. Both of manufacturing and agriculture industries which would be accompanied by the creation of new types of work not foreseen at the time.

Secondly, for business, the performance benefits of automation are relatively clear. Thus, the businessmen have opportunities for their micro economies to benefits from the productivity growth potential and macro economies to benefit to encourage continued progress and innovation , investment and market incentives. At the same time, employers must innovate policies to help workers and institutions adapt to the impact on

necessaries.

However, some authors agree (AI) will bring negative outcomes of automation, due to raise competiveness, reduce human job nature. Otherwise, other authors argue (AI) will bring positive outcomes of automation, due to raise productivities, job creation, assist humans work.

On the positive outcome hand, robots can increase productivity . This is particularly important for small-to medium sized businesses both are in developed and developing countries economies. It also enables large companies to increase their competitiveness through faster product development and delivery. Increased use of robot is also enabling companies in high cost countries to re shore, or bring back to their domestic base parts of the supply chain that will have previously outsourced to sources of cheaper labor. Currently , the greater threat to employment is not a automation, but an inability to remain competitive. Automation has led overall to an increase in labor demand and positive impact on wages. The reason is that the middle-income/middle-skilled jobs have reduced as a proportion of overall contribution to employment and earnings leading to fears of increasing income inequality, the skills range within the middle income bracket is large. Thus, robots are driving an increase in demand for workers at the higher -skilled and with a positive impact on wages. This issue is how to enable middle-income earners in the lower-income range to unskilled or retain. Finally, the (AI) positive impact supporter who argue the future will be robots and humans can work together.

However, on the negative outcome hand, robots can substitute labor activities, but don't replace jobs. They believe that less than 10% of jobs are fully automatable. Increasingly , robots are used to complement and augment labor activities, the net impact on jobs and the quality of work is positive. Automation can provide the opportunity for humans to focus on higher-skilled, higher-quality and higher-paid tasks. Robots can improve productivity when they are applied to tasks that which perform more efficiently and to a higher and more consistent level of quality than humans. For example, increased productivity is enabling some firms, such as Whirlpool, Caterpillar and Ford Motors company in the US restructure their supply chains, bringing back parts of the manufacturing process to the country of origin. Thus, productivity gains due to robotics and automation are important not just at the company level, but also for build industry and nation competitiveness.

I suppose that productivity can be raised. What are the impacts of robots on employment? Firstly, the main focus of development has been on personal entertainment, which does not drive worker productivity (manufacturing production). When the internet (information and communication technology (ICT)) innovation. This is borne and by findings that manufacturing productivity, which has been driven by innovations in automation rather than consumer technologies, has government strongly than productivity in the services sectors of the economy in most nature economies. It seems (AI) automation will create many jobs in internet communication entertainment game industry. For example, many young people like to use internet to play any electronic games from computer or mobile at home or outside home conveniently. Thus, (AI) automation will increase demand to be invented to any new entertainment game from internet channel. It will need to employ many (AI) entertainment game inventors to create many automation entertainment games. Thus, (AI) automation in internet entertainment game industry will need human (AI) entertainment game inventors to invent the knowledge-based capital of (AI) automation entertainment games. The (AI) entertainment game inventors will need own research and development skills, form specific skills, organizational know-how skills, databased knowledge, design and various forms of intellectual property to do these (AI) automation entertainment game invention occupations in the future.

International Federation Of Robotics(2016) indicated that China will be as a major robotics manufacturer and user of robots, benefiting from jobs created by robot manufacturing and productivity gains from robot use. Chins had sold of robots to any one single market every year since 2017 year. The Chinese government has included a focus on robotics in its 10 year strategy. In order to achieve its target of a robot density of 150 units per 10, 000 workers by 2020 year. Thus, Chinese companies will have to install around 650,000 new industrial robots between 2016 to 2020 year, 2.5 times more than installed globally in 2015 year.

Hence, China (AI) manufacturing industry will need to employ many workers . It implies (AI) manufacturing industry will create many new occupations in China. Also, ministry of economy, trade and industry (2015) also showed that Japan currently has the largest stock of industrial robots in operations, primarily in the automation industry. Driven by a rapidly aging population and low productivity rates, the Japanese government has sights on a 20-fold increase in the use of robots in the non-manufacturing sector and a three-fold growth rate of labor productivity in the service sector both by 2020 year. Thus, it also implies Japan will need many robots to be provide to service industry. Due to robots will provide to serve any businessmen's clients. Thus, it is possible that the service workers won't be dismissed as well as it is depended on the serving job nature to decide whether Japan's service workers can still serve to their employer when the service (AI) robots are applied to whose employers.

Consequently, it seems that (AI) can create employment, Ministry of economy, trade and industry (2015) showed that such as China will develop the major (AI) automation manufacturing industry. The (AI) employers will need to employ many workers to manufacture any these different kinds of (AI) robots to satisfy China or overseas individual or business buyers needs. But, (AI) can also cause unemployment to the low skillful service workers. Such as if Japan some service businesses choose to buy any (AI) service robots to replace their service staffs to serve their clients. It is possible that the service staffs will be dismissed, due to (AI) robots can do such as their same service job duties to achieve better service performance. Thus, today, it is increasingly common for people to use robots in various situations at home and in retail stores, hotels and hospitals these service industries. Robots are classified into server types based on their functionality (service and utility robots or those designed to communicate with humans) and appearance (humanoid robots or mechanical robots). The type of robot, to which each country allocated particular importance in the advance of robotics, reflects the sense of values and preferences of its population. Thus, if the country has high population needs to use robots, then they will influence either more new jobs creation or more old job loss in the country's (AI) manufacturing or (AI) service industries both. For example, Japan respondents often associate the term " robot " with humanoid robots that can communicate with human and they have a high level of familiarity with robot. The US has the highest level of robot utilization at home and in retail stores with its people being the most enthusiastic about the future use of robots. Germany shows a strong tendency to consider robots for industrial purposes and its people feel strong effort to the presence of robots in their households.

In conclusion, to judge whether how (AI) will influence the country's employment to be better or worse. It will depend on the country home buyers (users) or business buyers (users) how to use (AI) for their daily needs. If the country , such as US retail stores need to use (AI) , it will have possible to reduce some or many retail service workers. Even, if the country

, such as Japan has many home users need to use (AI) , it will not influence the employment market. Otherwise, it will raise (AI) salespeople numbers. Even, if the country, such as Germany and China will have many (AI) manufacturers, then it will create many (AI) manufacturing occupations for these (AI) manufactory workers.

Consequently, (AI) robots manufacturing and service needs will have positive or negative impact to any country's employment. It will depend on the (AI) service provision and service workers' job nature as well as the manufacturing workers of (AI) knowledge level to decide their employment chance in their country's employment market.

What does artificial intelligence(AI) development stage mean

● What (AI) function is?

Some scientists explain that artificial intelligence means which is an expert system, computer software that embodies a portion of the specialized knowledge of a human portion in a specific, narrow domain, owns decision making ability of human expert. The (AI)technology is based on the premise that what makes a person an expert is years of experience that enables who recognizes certain patterns in a problem as being similar to pattern. For example, in the future artificial intelligence system can be applied to control air traffic, design to computer configuration, medical diagnosis, instruction/training, speech/interpretation, monitoring to (nuclear plant), planning to mission, factory scheduling, prediction weather, repairing telephone, automatic driving etc. different industries.

Artificial intelligence characteristics include: creative, adaptive , common sense, fact processing, quick replication, broad focus permanent and consistent skill. Otherwise, traditional computer expert system disadvantage includes perishable, unpredictable, slow reproduction, expensive, slow reproduction, slow processing lacks inspiration, needs instruction, narrow focus only machine knowledge. So, artificial intelligence is a branch of computer science devoted to creating computer to influence software and hardware to attempt to create human intelligence or human intelligent behavior. It is learning from experience, responds flexibility in situation that are, new or not anticipated.

Thus, (AI) can be learnt programmed knowledge to solve problems, using reasoning in solving problem, understanding and inferring facts and rules, recognizing the relative importance of different elements in a situation. In summary, artificial intelligence is concerned with two basic ideas mainly:

The first idea, it involves studying the thought processes of humans to understand what intelligence is; the second idea, it deals with representing thought processes using companies to create artificially intelligent entities for testing the theories of intelligence.

● Can (AI) impact human job nature?

Human need concern this question: Will artificial intelligence (AI) reduce some human jobs in order to instead of replacing machines to do? Due to artificial intelligence is the ability of machines to do thing, that people would require intelligence. For example, artificial intelligence machine man driving(self-driver), it (AI) machine man driving research is an attempt to discover and describe aspects of human intelligence that can be simulated by driving machine functions. Alternatively, (AI) mathematical research may be another viewed as an attempt to develop a mathematical theory function to describe the abilities and actions of things (natural or man-made) exhibiting intelligent behavior and server as a design of intelligent calculation machine function.

Why do humans need artificial intelligence machines to instead of traditional human service job? For example, can artificial intelligence machine man (self-driving) driver drive to replace human driver? I shall compare the differences between humans and computers : The characteristics of humans are good at recognizing various things, either seen before or not, recognizing the relationship patterns between things. Human thinking is common sense reasoning, combining all types of sensory input, acting appropriately in novel situations, learning new things and changing behavior patterns, making decisions , even when given incomplete information, working with noisy, incomplete information gathering behaviors . However, characteristics of computers are good at: The tasks humans do naturally are extremely difficult for a computer program as intelligent, which must be able to do the same kind of tack as humans do naturally.

Hence, (AI) is an combination of many different success and technologies: Linguistics - computational and socio, philosophy-logic, philosophy of mind and of language, electronical engineering -image and speech processing, pattern recognition, robotics, machine learning, neural networks, optimization scheduling, management information system and decision making. So, it is possible that (AI) can impact human job nature to instead of human working behavior in the future.

● How can human society job nature
to be changed to artificial intelligent society?

From the first intelligent perspective reason view point, artificial intelligence is making machines " intelligent" acting as humans expect people to act. Artificial intelligence has ability to distinguish computer responses from human responses, it owns knowledge to solve expert problem. From another research perspective reason view point, artificial intelligence is the study of how to make computers do things which, at the moment, people do better (Rich & Knight, 1991, p.3).

(AI) researchers are native in a variety of domains, e.g. formal tasks (mathematics, games), tasks (perception, robotics, natural language, common sense reasoning), expert tasks (financial analysis, medical diagnostics, engineering, scientific analysis and other areas).

From the second business perspective reason view point, (AI) is a set of many powerful tools, and methodologies for using those tools to solve business problems. From a programming perspective reason view point, (AI) includes the study of symbolic programming problem solving and search .

From the third human technological perspective reason view point, today's computer can do many well-defined tasks, for example, arithmetic operations, are much faster and more accurate than human beings. However, the computers' interaction with their environment is not very sophisticated yet. How can human test whether a computer has reached the general intelligence level of a human being? Can a computer convince a human interrogator that it is a human? But before thinking of such advanced kinds of machines, human will start developing our own extremely simple " intelligent" machines.

So, it is possible that human society job nature will to be changed to artificial intelligent society when (AI) technology is developed to the mature stage in the future.

● Why does human need artificial intelligence machines?

One of major division in (AI) is between humans who think (AI) is the only serious way of finding out how we (human) work and human who want companies to do very smart things, independently of how we (human) work. This is the important distinction between cognitive scientists vs engineers. One of another major division in (AI) is between symbolic (AI), which represents information through symbols and their relationships. Specific Algorithms are used to process these symbols to solve problems or deduce new knowledge and connectionist. So (AI) , which represents information in network. Biological processes underlying learning, task performance and problem solving are imitated from human mind behaviors.

Thus, it is possible that artificial intelligence machines can do the better judgicious behavior to compare human.

● How does artificial intelligence influence future working changing in automation employment and productivity aspects?

In the automation changing influence aspect, as companies increasingly use robots on production lines or algorithms to optimize their logistics manage inventory, any carry out other core business functions. Technological advances are creating a new automation age in which ever-smarter and more flexible machines will be deployed on an ever larger scale in the marketplace. However, researching artificial intelligence with how influences human working nature. We need to answer these questions: How will automation transform the workplace? What will the implications for employment? And what is likely to be its impact both on productivity in the global economy and on employment?

Advances in robotics, artificial intelligence, and machine learning are growing in a new age of automation as machines match or outperform human performance in a range of work activities, including ones requiring cognitive capabilities. What factors are determined the changing in workplace adoption by artificial intelligence innovation? What advantages are automation? Automation of activities can be enabled businesses to improve performance by reducing errors and improving quality and speed, and achieving outcomes that go beyond human capabilities.

Some scientists indicated based on their scenario modeling. They estimated automation could raise producing growth globally by 0.8 to 1.4 percent annually. Almost, the activities people are paid almost $16 trillion in wages to do in global economy have the potential to be automated by adopting currently demonstrated technology. According to their analysis of more than 2,000 work activities across 800 occupations. When less than 5% of all occupations have of least 30% of activities that could be automated. They also indicated that technical economic and social factors will determine automation. Continued technical progress, for example, in areas such as natural language processing is a key factor beyond technical feasibility , the cost of technology, competition with labor including skills, and supply and demand dynamics, performance benefits including and beyond labor cost savings and social and regulatory acceptance will affect (alter) the scope of automation.

Other some scientists also indicate U.S. country for example, the anticipate shift in the activities in labor force of a similar order of magnitude as the long term sight away from agriculture and decreases in manufacturing. Share of employment in the United States both which were achieved. So, those factors can influence why artificial intelligence technology needs. So, it is possible that future agriculture and manufacturing both industries will apply (AI) technology manufacturer-kind of job nature to raise productivity instead of farmers, fruit picking workers, farming transportation labours as well as factory manufacturing workers and supervisors etc. human-kind of job nature.

● Is artificial intelligence possible to replace labor ?

Not just intelligence, but also debating, if machines are capable of having a conscious minds. Artificial intelligence has those characteristics as below:

On functionalism aspect, artificial intelligence inputs mental states, sensory inputs, (beliefs, desires being in pain feeling) and behavioral outputs. Since mental states are identified by a functional role, which are thoughts to be manifested in various systems. Even, perhaps computers which are physical devices with electronic substrate that inform computations on inputs to give outputs similar to brains which are artificial intelligence composed of part any intrinsic relationship to each other. Thus, artificial intelligence activities is not the whole itself, but into parts or on external influence on the parts.

On dualism aspect, artificial intelligence is a set of views about the relationship between mind are matter. On materialism aspect, it builds the only thing that exists is matter, including consciousness.

On biological naturalism aspect, it is similar a human brain than feels pains makes mental situation. So, artificial intelligence is similar biologist which might to be excited to human labor work.

Hence, it seems artificial intelligence can change (alter) or replace human labor work of nature in possible in the future.

● Can (AI) technology replace human labour nature of work?

On technological innovation reason view point, the history development of artificial intelligence studying the intelligence is one of most ancient scientific discipline. The history development of artificial intelligence what aims to achieve human use to sense, learn remember and think, logic probability, decision making and calculation develop from mathematics, instead of replacement human labor functions.

Artificial intelligence history development aim is the scientific analysis of skills in connection and practice with the appearance of computers from 1950 year beginning. The artificial intelligence (AI) can deal with the ultimate challenges. How can (either biological or electronic) mind sense, understand and manipulate a world that is much simple and more complex than itself? And what if would human like to construct something with such capabilities?

The general-purpose software of the early period of (AI) were only able to solve simple tasks effectively and failed when which should be used in a wider range or an more difficult tasks. One of the sources of difficulty was that early software had very few or mix knowledge about the problems which handled, and activities successes by simply syntactic manipulation. Moreover, the other difficulty was that many problems that were tried to solve by the (AI) were untreatable.

The early (AI) software whether trying step sequences based on the basic facts about the problem that should be solved, experimented with different combinations till which found a solution. From the end the 1960 year, developing the so-called expert systems were emphasized. These systems had (sue-based) knowledge base about the field which handled. Till to the beginning of the 1970 year, (Prolog) the logical programming language was born, which was built in the computation realization of a version of the resolution calculus. (Prolog) is a remarkably prevalent tool in developing expert systems (on medical, judiciary and other scopes), but natural language parsers were implemented in this language. Then, in 1981 s, the Japanese announced the fifth generation computer system project, a 10 years plan to build an intelligent computer system that use the (Prolog) language as a machine code. Nowadays, (AI) can be applied any industries, such as car manufacturing industry can use (AI) technological machine-men manufacture car, instead of replacing human labors in factory. Even, in the future, using (AI) machine-men drivers can drive any private cars or public transportation tools, instead of replacing human drivers, e.g. bus, train, tram, ferry etc. Also in the future, machine-men can replace housewives to serve families to do housekeeping clean job , e.g. cleaning toilets, bathrooms, kitchens, even cooking functions at home. So (AI) machine-man can reduce housewives works at home. Moreover, (AI) machine man can take care old people , when who are living at homes or elder care centers.

So, it seems artificial intelligence (AI) will be possible developed to manufacture a new generation machine-man to assist (serve) families to do any simply cleaning or cooking jobs at homes. Moreover, the overall demand of (AI) general social needs will also rise, such as security, driving transportation tools, restaurant cleaning, elder centers care service etc. So, it seems that individual or families or social needs of (AI) will be increase in the future. Thus, it will influence macro economy growth (GDP) if there are large house family consumer group and hotel or bus or taxis or ferry etc. different business consumer group demand any artificial intelligence machine numbers increasing. Then, the artificial intelligence products and material manufacturers must need to buy many artificaial intelligence materials to produce any kinds of artificial intelligence machines to prepare to satisfy consumer individual needs. Consequently, macro economy will grow to the owned artificial intelligence development countries, e.g. US, China, UK.

● Why can artificial intelligence satisfy human needs?

First, On machine-man satisfactory demand aspect view point, it makes computers that think, it is the automation of activities. We associate with human thinking: like decision making, learning. It is the act of creating machine that perform function that require intelligence when performed by people. It is the study of mental faculties through the use of computational models. It is the study of computations that make it possible to perceive, reason and act. It is a branch of computer science that is concerned with the automation of intelligent behavior. It is anything in computing service that human don't yet know how to do property.

Second, on thought aspect artificial intelligence means systems thank think like humans, systems that think rationally.

Third, on behavioral aspect, artificial intelligence systems that act like human and that systems act rationally. However, the basic objective of (AI) is to represent human's thought processes in computation . These machines are supposed to exhibit behavior that. It is performed by a human being, would be considered intelligent. However, some authors feel (AI) has disadvantages, such as it is not creative, it is excited in the use of sensory devices, it can't make use of a very wide context of experiences and it does not use common sense.

For speech recognition and understanding function needs example, (AI) can be applied in speech recognition and understanding function, which (AI) speech or voice recognition is a data input method. For example, the computer recognizes and understands one (or a few) word commands. Speech understanding on the other hand is the computer's ability to understanding a spoken language. That is , the computer understands the meaning of sentences, an paragraphs through (AI).

So, (AI) can be attempted to learn human language how to speak. It is similar to translate human language skill, instead of actual human speaking skill. Also, (AI) can assist handicap learning or language student how to listen different languages by machine-man sounds from computers more accurately.

So, it seems that it (AI) can replace human language teachers speaking function and can change teaching language nature of job in language speaking and listening education industry.

● Is artificial intelligence one good choice for human future technological benefit?

Nowadays, new technology development is popular. However, artificial intelligence is one kind of new technology choice among different technologies innovation. So it brings this question: Is artificial intelligence technology value to invest? To answer this question. I shall indicate some other new technology developments to compare (AI) technology development to judge which has urgent needs to achieve human expectation nowadays.

For example, why is green peace interested in new technologies? New technologies features prominently in our ongoing campaigns against genetic modified crops and number power. However, which are also an integral part of our solutions to environmental challenges, including renewable energy technologies, such as solar, wind and wave (water) power energy as well as waste treatment technologies, such as mechanical, biological treatment.

It seems humans need concern how to apply (AI) technology to solve environment pollution challenges in our future. So, environment protective, agriculture, natural energy technology will be popular demand to attempt to apply (AI) technology to solve their challenges or apply (AI) to assist to develop their industry.

How AI development stage to influence
economy growth

How can artificial intelligence technology influence economy?

Advances in artificial intelligence (AI) technology and related fields have opened up new markets and new opportunities progress in critical areas, such as health, education, energy, economic development, social welfare and the environment pollution.

(AI) automation will continue to create wealth and expand the global economy development in the future. However, when many will benefits that growth won't be costless and will be accompanied by changes in the skills, that workers need to increase productivity in the economy and structural changes in the economy. So, in the skills that workers need to succeed in the economy and structural changes.

I shall indicate why aggressive policy action will be needed to help Americans who are disadvantaged by these changes , due to (AI) technology is caused. For automation industry change example, artificial intelligence (AI) capabilities will enable automation of some tasks that have long required human labor. These artificial intelligence technology introduction can increase new opportunities for individuals. The economy and society, but (AI) has also the potential to disrupt be current livelihoods of many Americans. However, (AI) leads to unemployment and increase in inequality over the long run depends not only on the (AI) technology itself, but also on the institutions and policies that are changed.

Thus, it is possible that (AI) technology will raise some countries unemployment number if the employer apply (AI) technology workers to work instead of human labor in their factories, but it can also raise productivities for these employers.

● Can (AI) influence global economy growth?

Technological progress is main driver of growth of GDP per capita, allowing output to increase faster than labor and capital . However, technology can increase productivity, but also decrease the number of labor hours needed to create a unit of output. So (AI) causes unequal to labor wage decreases, even reduces the number of labor to manufacture, e.g. artificial intelligence technology of automation car manufacturing industry; clothing manufacturing industry; plane manufacturing etc. high technology of artificial intelligence manufacturing method. But (AI) should be potential environment benefit, although it raises unemployment ratio. Moreover, it can rise production , due to many skilled craft were replaced by the combination of machines and lower-skilled labor. The result of (AI) technology introduction , it causes output per hour risen when inequality declined, driving up average living standards, but the labor of some high-skill workers was no longer as valuable in the market. Otherwise, if (AI) technology is continue developed to be success. Some routine intensive occupations will be loss, which focused on predictable, e.g. easily programmable tasks, such as switchboard operators, filing clerks, travel agents, and assembly line workers would be particularly replaced by new (AI) technology. However, at the same time, (AI) technology development will bring these benefits: improvement in education (training (AI) technology scientists) , due to (AI) manufacturing technology needs are raising to businesses and institutional changes, such as the reduction in unionization and raising in the minimum wage to the (AI) manufacturing technology skilled labor in factories.

Because (AI) technology is not a single technology, but rather a collection of technologies that are applied to specific tasks, the effects of (AI) will be felt unevenly though the economy. It will bring some tasks will be most easily automated than others , and some jobs will be affected more than others, both negatively and positively. Finally, new jobs are likely to be directly created in areas , such as the development and supervision of (AI) as well as indirectly created in a range areas though out the economy as higher incomes lead to expanded demand.

However, if (AI) technology could dominate global labor markets. If labor productivity increases, do not influence into wage increases, then the large economic gains brought about by (AI) technology could be increased wealth inequality, due to employers can reduce production cost, but workers (labors) wages will not be increased, even will be decreased. Hence, it seems the (AI) technology will bring disadvantages to labor market to cause unemployment or reduce wages in possible, although it can reduce employer individual salary (wage) expenditure and it can raise productivity.

● How can artificial intelligence impact global economy growth?
Artificial intelligence (AI) technology is a branch of computer science that aims to create intelligent machines that work and react like humans. So, (AI) is a technology that appears to impact (influence) human preference by learning, understanding complex contents, enhancing humans in executing both routine and non-routine tasks. In the future, (AI) technology that can be virtual personal assistant, as well as it may exist, such as robots with human-like processing capabilities.

How can (AI) technology impact global economy growth over the next 10 years? During this time period, (AI) technology is predicted to have wide-ranging applications including: Machine learning that automates analytical model building by using algorithms that allow machines to operate without human assistance.

In global education aspect, potential applications include predicting cause-and-effect relationships from biological data, identifying new drugs, self-driving cars, and protecting against fraud, improved natural language processing that allows computers to continue to better analysis, understand and generate language to interface with humans using natural human languages. For example, transcribing notes dictated by physicians, automatically drafting articles and translating text and speech. So (AI) technology can be applied to education aspect to improve humans' knowledge level.

In visual art aspect, (AI) machine vision that allows computers to identify objects, scenes and activities in images. Current applications of (AI) machine vision include providing objective descriptions for the blind seeing(visual) needs.

We except the economic effects of (AI) technology to include both direct GDP growth from sectors that develop or manufacture. (AI) technology and indirect GDP growth through increased productivity in existing sectors that employ some form of (AI). If (AI) technology is an increasingly critical component of more products, it will become an integral part of many people's lives. Thus, (AI)'s ability to influence economic activity, rather than the economic or development status of the region. (AI) has the potential to impact income classes and to bring significant gains to both developed and developing countries. For example, (AI) has the potential to optimize good production around the world by analyzing agricultural regions and identifying what is necessary to improve crop yields.

In estimating the future economic effects by (AI) technology innovation, it is important to note that it is challenging to accurately predict which applications of (AI) will ultimately be commercially successful. In micro level economic influence, we need to apply methodologies to estimate the economic effects of investment in firms developing (AI) technology since investment levels in a technology are a telling sign of the future potential of that (AI) technology.

● How can (AI) influence GDP of high income countries in the next ten years?

How (AI)'s development may affect the global economy over the next ten years. In fact, (AI) technology has the potential to affect business across the global in a wide range of industries in ways only a number of technologies have done in the parts. For example, (AI) technology's expected to be a useful tool for enhancing human capabilities and in some instances replacing functions, such as driving a car, adoption of broadband internet, mobile telephone, industrial robotic automation have served to enhance human capabilities.

However, significant public debate has focused on projections of (AI) technology's effect on the labor force. However, large companies prefer to invest in (AI) technological industry. For example, face book's (AI) research lab., google machine intelligence lab. and micro soft machine learning and artificial intelligence research division are all making advances in (AI) technology and investing in the industry's top talent. Additionally, between 2010 year and 2015 year, nearly $5 billion in venture capital funding invested in firms across the global developing and employing (AI) technology (Facebook (AI) Research).

● How can artificial intelligence impact on workplace?

Modern information technologies and the labor economy growth of machines is powered by artificial intelligence have already strongly influenced the world of work in the 21 ST century. Computers, algorithms and software simplify every tasks and it is impossible to image how most of our life could be managed without them. How can be the information economy characterized by exponential growth replaces the most production industry based on economy of scales? What will the future world of work look like and how long will it take to get? Will the future world of work be a world where humans spend less time earning their livelihood? Alternatively, are mass unemployment, mass poverty and social distortions also possible scenario for the future, where robots, artificial intelligence systems play an increasingly central role? These questions concern how artificial intelligence further development . Can influence labor economy growth on workplace ? When the labor market has widespread impact on intelligence property, information technology, product liability, competition and labor and employment laws.

How (AI) technology impacts on labor workplace.

The future influence any organizations how labor economies use of (AI) can be analyzed, such as deep machine learning is based on a set of model high level data. Unlike human workers, the machines are connected the whole time in workplace. If one machine makes a mistake, all autonomous systems will keep this in mind and will avoid the same mistake the next time.

Over the long run intelligent machines will win against every human expert.

Production robots have been replacing employees because of the (AI) technology. They work more precisely than humans and cost loss. Creative solutions like 3D printers and the self learning ability of these production robots will replace human workers, the automatic data recording and data processing, traditional back office activities are no longer in demand. Autonomous software will collect necessary information and will send it to the employee who needs it. Additionally, dematerialization leads to the phenomenon that traditional physical products are becoming software. For example, CD or DVDs are being replaced by streaming services. The replacement of traditional event ticket, e-travel ticket service products or hard cash will be the next step, due to the possibility of payment by smartphone. So, (AI) technology will impact human's daily life consumption behaviors in the future. For another example, transportation tools, such as boats and ferries and private vehicles will use sensors and navigating without human input. Taxi and truck drivers will become obsolete, the stock store applies to stock managers and postal carriers of the delivery is distributed by (AI) machine delivery method.

What is the relationship between (AI) and (CRM)?

● Can (AI) technology impact on customer relationship management (CRM) ?

Nowadays , (AI) is a technology almost as old as the computer industry itself, it is similar with the advent of personal assistants function to businesses and personal promotion channel, such as (Amazon's Alexa, Apple's Siri, Google's Assistant) image recognition (face book), personalized recommendations (Netflix , Amazon). Those innovations have been driven by a increase in processing power, lower cost hardware, and the exploding creation and availability of data. It seems, (AI) technology can impact global customer service management method.

How to forecast economic impact modeling to (AI) will affect global economy? Can human forecast business revenue growth and job creation (or destruction) based on (AI) applied to customer relationship management (CRM) activities? In addition to the economic impact on (AI) or (CRM) which can include an estimate of the economic impact attributable to sales forces customer base. What can economic benefits be brought to (CRM) from (AI) technology?

Artificial intelligence(AI) comprises a set of technologies that use natural language processing, machine learning, knowledge graphs, and other tools to answer questions, discover insights and provide recommendations. Computer systems can use (AI) hypothesize and formulate possible answers based on available evidence can be trained through the ingestion of vast amounts of content, and automatically adapt and learn from (AI) self mistakes and failures.

So, any business organizations (customer service departments) can provide efficient and effective customer relationship management of excellent customer service quality if which applied (AI) technology system. The different type of (AI) systems include: (AI) system platforms, machine learning (AI) based data preparation and enrichment tools, machine vision/image recognition, voice speech recognition, text analysis and natural language processing, bots , e.g. face book website and virtual digital assistance solutions, social media pattern analysis , sentiment analysis, advanced numerical analysis (e.g. IOT streaming , machine logs), supporting technologies, knowledge base dialog management, Q&A processing etc. different (AI) technology system customer relationship management (CRM) tools.

(AI) (CRM) of activity can include these categories, such as: corporate marketing, marketing operation, field marketing, customer support, digital commerce, customer analytics, customer influenced product or service design, product or service pricing, finance information, presentation, customer billing, inventory , logistics and fulfilment support, partner management etc. different CRM tools.

(AI) technology of CRM has been carrying on plan different stages to achieve CRM personal assistant tool for businesses. The stages are such as, in the beginning stage of (AI) projects in place, implement now, pilot phase next year in the final stage of (AI) customer relationship management tools are foreseeable future. So, this CRM technology has been improved to plan in different stages every year to prepare to achieve full capacity of CRM service quality for businesses to use in the future.

Hence, how to develop an estimate prediction of the economic impact (AI) technologies could have CRM activities, which depends on gathering macroeconomic information on business revenue and the basic marketing of business revenue and the basic markup of business expenses by major functions (customer support, marketing and sales , production etc.)

An economic impact model that can gather data together and forecast the results how (AI) artificial intelligence technology brings (CRM) customer relationship management benefits to businesses, e.g. surveys investigation includes IT spending by sample countries, GDP and population estimates and forecasts, revenue per employee and ratios of IT spend to GDP. Surveys (questionnaire questions) of forecast results are influenced by (AI) impact can include: results are projected from surveys and rely on estimates are made by respondents on the expected financial improvements in categories of (AI) –assisted customer relationship management activities. The forecast assumes that these estimates are correct; financial estimates are based on estimates of "first year" improvement from full (AI) implementation; forecasts are from planning to implement any artificial intelligence of customer relationship management (CRM) projects, the improvement forecast is of categories of activity , e.g. corporate marketing , digital commerce, and customer analytics. They are not estimates of ROI for the (AI) software. They rely on conservative estimates to which each of these entities might affect company revenue, expenses or productivity. They also rely on estimates of the penetration of software in customer relationship management activities . Net new jobs created are based on the ratio of new revenue to jobs required to support that revenue . They can assume that 50% of the net new revenue will support increases in labor and the rest will go for capital and other operating expenses that may replace jobs lost to automation.

In the future, some of the ways in micro economic benefits to any organizations. (AI) technology is expected to impact CRM activities include: Spending up sales cycles, improving lead generation and qualification solving customer support problems faster (raising service quality), helping companies improve brand campaigns and recognition, lowering costs of support calls when increasing resolution rates, lowering the cost of recruiting employees and partners, increasing revenue from optimized product marketing, optimizing price, distribution logistics and preventing loss through fraud detection. So, micro economic benefits view point, it seems that (AI) CRM technology can raise any companies economic benefits for care term.

Artificial intelligence enables machines or the in-build software to behave like human beings which allows these decisions and act. The advent of (AI) is leading , talking, making decisions and act. The advent of (AI) is leading to new technologies advances and transforming the economic and employment opportunities for humans in a positive way. (AI) related technologies can facilitate our live. For example, industrial robotics, robotic medical assistants, smart games, financial forecasting software, big data analysis, algorithms in health and bioinformatics, pilotless cargo places, drone ambulances and general purpose and workplace robots and others. (Disruptors technologies: Advances that will transform life, business and the global economy).

Artificial intelligence also known as computational intelligence is defined as " the human –like intelligence exhibited by machines or software. It is theorized that intelligence of humans can be described and intelligence machines or software can simulate it. These machines software can be reasonable , learn, perceive and process information, like human mind and thus facilitate human life. They can think and act for us. So, artificial intelligence is an interdisciplinary field of study including computer science, neuroscience, psychology, linguistics and philosophy.

However, (AI) research and developments have economically impacted many industries, such as robotics, telecommunications, computer applications , health, finance, heavy manufacturing, transportation, aviation, e-service and e-commerce, military , music and movie, toys and games entertainment etc. industries.

In fact, many ideas, systems and technologies have been developing in the world of (AI) technology. However, which are net called or considered (AI) products, rather which are mentioned with their specific names, such as smart graphics, machine learning, e-commerce etc. (i.e. this is called (AI) effect).

● How can (AI) technology influence digital economy?

Nowadays, (AI) related industrial applications will replace most human power in fields, including call centers, customer services and air cargo transportation. (AI) technologies also help weather forecasting based on repeated rainfall pattern (data) recognition, through robotics (i.e. floor cleaning, moving lawns etc.) transporting people and products with unmanned vehicles, sending space unmanned smart shuttles, developing robotic arms, predicting market values in stock exchanges by internet, making homes safer, helping elderly and disabled using robotic servants etc.

Among the (AI) related technologies , there are a few that significance for the impact on society and especially on digital economy . (AI) is particularly influential in machine learning. Such as robotics, transportation, finance, health and bioinformatics, e-commerce , e-games, big online data gathering and internet-of-things. For example, machine e-learning is based in bioinformatics and robots that can learn new skills for better caregiving in healthcare. What is machine e-learning? Machines can e-learn from e-data gathering, coming up generalizations and making decisions to act in certain ways from internet.

There are important applications , such as e-machine perception, electronic online natural language learning processing, online search engines, online bioinformatics, online brain –computer interface, online game playing, online robot locomotion, online advertising, online computations finances, online health monitoring, online DNA classification and decision making, online in chemistry –cheminformatics . So, online machine learning can positively impact productivity and it can enhance information and analytical system from (AI) online channel.

What is robotics? Robotics is one of the most strongly influenced fields in (AI). For example, heavy manufacturing industries, robots and used and man power is replaced for effectiveness, precision, and accuracy, especially in respective or dangerous tasks, including welding, assembling , picking and placing .

So, robots can acquire new skills or adapt the changing dynamic environment. Also, artificial intelligence can be applied in developing transportation. For example, automated vehicles, driver assistance systems , safety systems, collision avoidance systems and public transportation. Moreover, (AI) technology has proven to produce some of the best tools to predict stock market fluctuations from internet data gathering method. It's predictions are based on ever-evolving predictions algorithms and systems learn new models and make connections between historical data and new data to measure stock market trading more accurate from internet data gathering channel.

In health field, especially in health data processing , analysis, decision making support and medical diagnosis. So, online data can show which patients will need what treatment and what alternative drugs could be used more accurate from (AI) online data gathering method. Bioinformatics is an interdisciplinary field combining statistics, (AI) online technology can help in discovering data patterns and modeling through the application of machine learning, artificial neural networks and genetic algorithms. For example, further (AI) technology development of human genome project of online data sequences.

Online shopping can be facilitated by virtual assistants developed through (AI) technology and these assistants can offer the best advice. (AI) online purchase coming after every product image recommendations and personalization bring important revenue to shopping online sites, like Amazon . Smart computer graphics and games, artificial intelligence is useful in smarter computer, graphics, scene modeling , scene rendering processes in order to create, for example, effective human −robot interactions , online machine learning, online strategic games techniques etc. online computer related (AI) software.

So, online big data analysis and big data does have a critical need in the world of online intelligence machines and software in our future. In other words, (AI) offers online technology to enable online big data analysis to provide industrial organizations with valuable information for effective decision making in short time. For example, what IBM's Watson achieved: this machine used 200 million of structured and unstructured content with a special technology of hypothesis generation, massive evidence gathering, analysis and scoring from internet channel.

Finally, (AI) online technology another related internet invention (internet of things) (IOT) is the network of machines or objects connected through internet. These connected objects can sense their internal and external environment, communicate with each other, can send critical data and finally can make decisions to act or correct their environment from (AI) online technology. For example, factories can monitor and automatically change production processes, hospitals can monitor and regulate the health conditions of their patients , schools can collect data from facilities and cars can send data to car makers from (AI) online technology.

Partner predicts that (IOT) market will create about trillion amount value by 2020 year. Although machines collect big data from their environment, whether which gain an insight or learn from these online data largely depends on the (AI) online machine learning principals and (AI) online technology. In 2013, Mckinsey estimated that disruptive technologies closely related with potential economic impact in 2025 year between $7.1 to $13.1 trillion amount (automation of knowledge work, advanced robotics, autonomous or near-autonomous vehicles).

What is the relationship between
(AI) and global digital economy development ?

● Could work activities in China be automated
making in the nation with the world's largest automation potential?
Can (AI) technology influence China economy? Could China workers be

affected and jobs made up of routine work activities and predictable? Will programmable tasks be particularly impact to China employment market ? When impact on labor market is likely to be gradual at the aggregate level, it can be sudden and dramatic at the level of specific work activities, rending some job obsolete fairly. Overall (AI) technology will raise digital skills when reducing demand for medium incomer inequality for China workers. It seems (AI) technology's effect on productivity could be crucial to China's future economic growth as the population ages are increasing.

In China, some biggest technological companies driving significant investments in research and development. Moreover, China is one of the leading global (AI) technology development county. However, China will need to focus on building its innovation capacity. For example, United States and United Kingdom are currently producing more influential (AI) technological research. However, if China planed to achieve (AI) technology success, it's traditional industries will need to develop technical know-how –to and overcoming implementation costs prepare to develop (AI) . When (AI) technology is introduced into China society, China government needs to raise concerning ethical, legal, technological security etc. business questions. Also, surrounding issues include privacy, discrimination, legal liability and regulation. It aims to encourage overseas investors to choose to invest (AI) technological industry to raise GDP growth and manufacturing industries income growth for long term in China. If China encouraged overseas (AI) technology investment in its country. It is possible to influence China employment market to be changed. Because (AI) technology will impact to influence China people daily life. Due to (AI) technology is introduced to China society, many rich people will prefer to spend to buy any high (AI) technological products for entertainment or learning or machine man driving etc. daily necessity activities. Then it will raise GDP growth and will raise (AI) manufacturers or related-(AI) technological manufacturers profit. It is beneficial to China because it can become one high knowledgeable and (AI) technological economical society. But it will bring bad influences to raise unemployment chance for the low skillful labor. In labor economy aspect influence , how (AI) technology can influence China low skillful labor unemployment ratio raising. The raising low skill labor unemployment reason is because China low skillful human labors are argued or are replaced by (AI) technology creating new challenges to introduce to influence China society of simply human manufacturing job nature to be changed to be high (AI) technology manufacturing job nature in any China factories. Moreover, when (AI) technology introduction to China, it will cause other related social challenges in China. The varied (AI) related challenges, including the difficulty of creating safe and reliable hardware for sensing and affecting (transportation and education), the challenges of gaining public trust, a low resource comities and public safety and security, the challenges of overcoming fears or marginalizing humans in China employment and workplace and the risk of diminishing interpersonal trust because the low skillful labors won't believe any China employers will give chance to employ them , due to (AI) technology will replace their skills and man manufacturing of productivity is much less to compare to (AI) technology manufacturing method.

● How does (AI) technology influence
the future of employment change?
Are future nature of jobs changed to computerization from (AI) technology? Where are the probability of computing occupations from (AI) technology influence? What is expected impacts of future computing on labor market from (AI) technology influence? John Maynard Keynes's frequently cited prediction of widespread technological unemployment " du to our discovery of means of economic the use of labor outrunning the pace of which we can find new used of labor" (Keynes, 1933, p.3).
In the future, (AI) technology will impact some nature of occupations to change computing. This chance will also influence some countries' economic change. For example, some factory human labors hand routine manufacturing tasks will be changed to computerization of routine manufacturing tasks by (AI) technological machine men hand manufacturing method. it will cause a structured shift in the labor market, with workers reallocating their labor supply from middle-income manufacturing to low-income service occupations.
Arguably, this is because the manual tasks of service occupations are less computerization, as who require a higher degree of flexibility and physical adaptability. So, (AI) technology will influence the human hand labor skillful occupation nature of task cheaper , such as vehicle manufacturing , ship manufacturing, computer manufacturing, steel manufacturing, television, radio etc. home electronic products of heavy machine industry change. Due to (AI) technology machine man will be proper to be used to manufacturing these electronic products when the (AI) technology innovation can develop to the mature stage. Then, any countries

manufacturers will choose to use (AI) technology machine man, instead of human hand production.

Supposing the future prices of computing are fallen, seriously, problem solving skills are becoming relatively productive, explaining the substantial employment growth in manufacturing occupations, involving cognitive tasks where skilled labor has a comparative advantage, as well as the increase education needs for (AI) technology computing of machine man subject study.

Prediction of education needs for (AI) technology student numbers will increase, due to manufacturing industry needs many (AI) technology students in future employment market. Another (AI) technology influence if the future (AI) technological innovation, e.g. machine man manufacturing or machine man service industries will both increase demand, then with more sophistic software technologies will be disrupted labor markets by marketing workers redundant.

For publishing industry, what is striking about the case in paper book publishing industry will be unpopular? Due to the electronic book publishing industry will be popular, e.g. Amazon publish . (AI) technology can influence paper book manufacturing method which is replaced by machine man electronic book manufacturing method as well as it will cause the computerization is no longer confined to routine manufacturing tasks. Due to (AI) machine man manufacturing technology will be proper to be used to manufacture any products in short time efficiently and effectively , e.g. electronic book products. In the future, if it is fact to occur this case, such as (AI) technological machine man manufacturing method will be adopted (applied) to manufacture electronic books or any products in possible. (AI) technology will cause many manufacturing workers are unemployed. It is beneficial to employers, who can reduce to spend much wages expenditure to employ manufacturing workers, but it will cause many manufacturing workers loss jobs and reduce income to support whose families lives. It will cause social challenges, e.g. increasing stealing crimes if the manufacturing workers had not other skills to find other jobs to do easily. So, manufacturers need to concern over technological unemployment which will be hardly future phenomenon if who decided to dismiss all manufacturing workers, due to (AI) technology machine men replace to them.

If (AI) technology can be innovated to produce any kinds of machine man to serve any service or manufacturing industries successfully. Then, it will bring these questions: Can future that workers be influenced to be automation employment and productivity by (AI) technology influence? Does it impact to influence the (AI) technology countries' productivity and growth and natural resources development and labor markets and evolution of global financial markets and economic impact of technology and innovation and urbanization etc. issues? How will automation transform the workplace? What will be the implication for employment? What is likely to be its impact both on productivity in the global economy and on employment?

In fact, automatic of activities can enable businesses to improve performance by reducing errors chance and improving quality and speed, and same cases achieving outcomes that go beyond human capabilities. Some economists indicate (AI) technology would give a needed boost to economic growth and prosperity have of the working age population in many countries. Based on the scenario modeling, they estimate automation could raise productivity growth globally by 0.8 to 1.4 % annually. They also indicated that almost half the activities people are almost $1.6 trillion in wages to do in the global economy have the potential to be automated adapting current demonstrates technology, according to their analysis of more than 2,000 work activities across 800 occupations. When less than 5% of all occupations can be automated entirely using demonstrated technology, about 60% of all occupations have at least 30% of worker made activities, that would be automated. More occupation will change to be automated. They also indicated for business performance benefits of automation are relatively clear, but the issues are more complicated by policy making to attract foreign investors. Beyond technical feasibility, the cost of technology, competition labor will include skills and supply and demand dynamics, performance benefits and beyond labor cost savings and social and regulatory acceptance will affect the automation. Their predictions suggest that half of today work activities could be automated by 2055 year, but this could happen 10 to 20 years earlier or latter depending on the various factors in addition to their wider economic condition.

Some scientists suggest (AI) technology is finally starting to deliver real-life business benefits. Computer power is growing significantly , algorithms are becoming more sophisticated and perhaps most important of all, the world is generating vast quantities of the fuel that powers (AI) technology data billions of gigabytes of it every day. Also, online firms are digital natives, such as Google online search service company is investing on (AI) technology. For new though most of the news if coming from the suppliers of (AI) technologies. And many new users are only in the experimental phase. Few products are on the market or are likely to arrive these soon to drive immediate and widespread adoption. As a result, analysts believe (AI) technology's potential will give true economic benefit in the future. (AI) industry will introduce to suppliers and users to raise economic potential of (AI) technology.

In the future, (AI) technology systems can solve business problems. Some scientists categorized those into five technology systems that are key areas of (AI) technology development: robotics and autonomous vehicles, computer vision language virtual agents and machine learning , which is based on algorithms that learn from data without replying on rules-based programming in order to draw conclusions or direct an action.

Such as computer vision and language includes natural language processing, analytics, speech recognition technology, some are about learning from information, such as about machine learning and others are related to acting on information, such as robotics, autonomous vehicles and virtual agents, which are computer programs that can converse with humans. Machine learning and a subfield called deep learning are artificial intelligence applications.

● Can artificial intelligence impact
global economy growth?

Artificial intelligence (AI) is a term first defined in 1956 year. It is a branch of computer science that aims to create intelligent machines that work and react like humans. In contrast today, 60 years later, (AI) is characterized by a number of applications, including computers playing games against humans and understanding human languages, virtual personal assistants, and robotics which involve computers seeing , hearing and reacting to sensory stimuli. In the future, technologists predict for (AI) technology ranging from (AI) being used as a tool to aid relatively simple processes for robots with human like mental capabilities, who expect (AI) technology can emulate human performance by learning, coming to mind its own conclusions, understanding complex content, engaging in dialog with people, enhancing human cognitive performance or replacing humans in executing both routine and non-routine tasks. In existing industry, (AI) technology is used , such as targeted advertising and virtual used personal assistant as well as the (AI) technology that my exist in the future, such as robots with human vehicle processing capabilities.

The range of (AI) technology's progress in the future will determine the economic impact future of (AI) technology on the global economy with more limited advances and applications (i.e. weak (AI) only) corresponding to more limited economic impacts and more substantial progress, i.e. strong (AI) technology is corresponding to more significant economic impact.

(AI) technology learning that automates analytical model, including predicting cause-and-effect relationship from biological data, identifying new drugs, self-driving cars and protecting against fraud etc. functions. Also (AI) learning can improve natural language processing that allows computers to continue to better analyze, understand and generate language to interface with human using the natural human language, virtual personal assistant, helps users by providing scheduling appointment, reminds organizing personal finance and finding providers of various services, machine vision allows (AI) machine man to identify object, scenes and activities in detect pedestrians and bicyclists.

We expect the economic effects of (AI) technology to include both direct GDP growth from sectors that develop or manufacture (AI) technology and indirect GDP growth through increased productivity in existing sectors that employ some from of (AI) technology. If (AI) producing sectors could grow, then it could lead to increase revenues and employment of (AI) technological professionals within these existing firms as well as the potential creation of entirely new economic activities to any countries' societies productivity improvement in existing sectors could be realized through faster and move efficient processes and decision making as well as increased (AI) technological knowledge and access to information available in societies easily.

In the future, if (AI) technology is an increasingly critical component of more products, it will become an integral part of necessary products of many people's lives. The extent of (AI)'s economy effort is also likely to vary from region to region, thought variation may be more dependent on the predominate economic activity of a region and the (AI) ability can influence economic activity, rather then the economic or developmental status of the regions. (AI) technology can move accessibility and can use source development to do international business between one country and another country.

So (AI) technology has the potential to give benefits to different income chooses and to bring significant gains to both developed and developing countries. For agricultural technology, (AI) has the potential to optimize food production around the world by analyzing agricultural regions and identifying what is necessary to improve crop yield. In total, (AI) technology gives greater economic impact to any countries agricultural regions if which implemented (AI) technology to grow crop , fruit etc. food production in the farms.

Investment in (AI) technology is such as capital investment to any countries' public or private enterprises. So, it will have large economic impact to the future . If the (AI) technology is reasonable invested to the different needs aspect by the public or private enterprises in the country. Then, it will have good economic impact to the country in the future. However, when (AI) technology is likely to affect both the productivity and employment components of economic growth in many sectors. Significant public debate has focused on projections of (AI)'s effect on the labor force. However, for instance, some researchers have argued that the rise of (AI) technology and automation will led to significant unemployment as capital is substituted for the low skillful labor. So, they point to the concern that the increasing sophistication of (AI) technology may balance skilled and semi-skilled workers and the reduce the size of the middle class. However, this is not a new argument, due to (AI) technology negatively affecting the labor force and leading to mass unemployment. Because the (AI) technology is the substitution of machinery for human labor. Although, employment in certain industries, has been reduced in the past due to technological advancement. For long term, the labor market has adapted to the introduction of new technology, giving rise to new jobs in new areas. (AI) technology may also be accomplished without a reduction to total employment in the long-term to some Asia countries, such as Hong Kong and Japan. Because Hong Kong and Japan many low skilled labor, e.g. security, cleaner who complaint that employers need them to work long time hours. (abnormal working hours) e.g. one day 12 to 15 working hour per day. Hence, if (AI) machine means invention technology success. Security or cleaning job can be worked from (AI) machine man in some hours every day in order to reduce the long time working hours cleaners or security workers, e.g. one (AI) machine man works 4 hours for cleaning or security job, one day as well as another cleaner or security labor only needs to work 8 hours one day. So total security or cleaning employers can employ 12 hours machine cleaners or security workers and human cleaners or security workers in one day. For long term benefit, Hong Kong or Japan every security or cleaning worker does not need to work 12 hours minimum working hours one day. They won't feel tried and bore and without private with whose families, so who will accept to do these cleaning or security jobs, even they can raise work efficient and performance when who feel happy and health.

So, (AI) technology of machine man invention can raise low skillful labor efficiency and it can help them to avoid abnormal working hours demand in some busy work life countries, such as Hong Kong and Japan. Before, one Japan female labor feel unhappy to work, due to who often needs to work abnormal working hours for her employer and who has less sleeping and without any private time to enjoy her life with her families every day. So this abnormal working hours factor causes her to do commit suicide behavior, then she is die unlucky. So (AI) technology of machine man invention ought avoid abnormal working hours demand for employer in any countries in the future.

The most important occurrence to any employers, some researchers had attempted to do one experiment to find that private research and development , venture capital and public research and development investment all have strong net effect or economic growth with venture capital funding further having the strongest such effect from (AI) technology. The researchers hypothesize the venture capital investment contributes to economic growth through (AI) technology innovation and by the capacity of an economy to use existing (AI) technology knowledge to increase productivity. They predict the impacts of venture capital, business-research and development and public research and development can raise multi factor productivity from (AI) technology introduction.

Can (AI) technology influence the economic development to developing countries? The developing regions of the world contain most of natural resources. If one day, (AI) technology has invent one kind of machine man which can assist any gas or oil workers to seek any new oil/gas natural resource locations easily. I believe that (AI) technology can help these natural resource exploitation countries will gain economic benefit more easily. So, (AI) driven technology can be used to change to create any new opportunities to address poor management or resources and improve human well being, such as Africa Latin America and India can use (AI) technology machine man to seek any oil/gas natural resource countries exploitation activities to attempt to gain much economic benefits.

● Why will (AI) technology grow economic
development ?

Nowadays, increases in capital and labor are no longer driving the levels of economic growth, such as (AI) technology. The ability of increase in capital investment and in labor of traditional drivers of production, have no longer to be enjoyed in most developed economies ,e.g. developed country, US, UK . However, artificial intelligence has the potential to overcome the physical limitation of capital and labor to avoid missing out on this opportunity. So, policy makers and business leaders must prepare for and work toward a future with artificial intelligence. They must do with the idea that (AI) is another simply method to enhance productivity method . Rather they must see (AI) as the tool that can transform thinking about how growth is created.

Economists have always thought of new technologies are as driving growth their ability to enhancing. It can replace labor and capital factor of production. So, it brings this question: What is the factor of production (AI) technology characteristics. They key factor is to see (AI) technology as a capital-labor .

(AI) can replicate labor activities at much greater scale and speed, and to even perform some tasks began the capabilities of human. For example, by using virtual assistants , 1000 legal documents can be reviewed in a matter of days instead of taking three people six moths to complete. Some (AI) technology may be one kind of factor of production in the future. For another example, people will work in workplace digitalization environment. So, in the future, working environment and information management are automated. Such as Konica camera sale company will use workplace digitalization. So , (AI) technology can provide workplace digitalization in order to raise productivity efficiency. (AI) technology will be one kind of production which is replaced by workplace digitalization and it will grow any organization productivity efficiently. Then, (AI) technology will assist overall social economy growth , due to productivity is raised and products can be produced in short time to prepare to sell in consumption market. So, time will be shortened to increase GDP growth fast for the development of (AI) technology countries.

● How can (AI) technology impact to global
economic and social and psychological

changes?

What will be the development of (AI) technology and predictions concerning the future evolution? The computers and robots will develop conscious, intelligent and minds into humans, enhancing psychological and behavioral abilities and allowing for direct communication with (AI) minds. (AI) technology will be impacted human life by (AI) technology information communicative and environmental influence. A " world brain" and " world mind", this psychological system will be enhanced and enriched the capacities of both individual and collective cognition by (AI) technology of service industries.

(AI) technology with influence these human needs of service industries changes, such as , biological science, finance, entertainment, business, biological science, transportation, communication military etc. The personal computer evolution, the internet and the world wide web which exploded on the scene, linking business, homes, schools, social organizations which were a completely unpredicted phenomenon to influence human life. Kurzweil (1999) predicts that by 2029 year, most human communication will be with machines. According to Person, by 2100 year, there will be human machine convergence.

How can (AI) technology influence environmental protection to make benefits to farming economic growth? (AI) technology can be applied to predict how to solve environmental pollution challenge to avoid to damage any crop or vegetable or rice or fruit etc. food growth. Because environmental experts can gather global environmental pollution data from an environmental database to build a perform a systematic analysis from (AI) technology. The first step is this broad analysis can include understanding, statistical and data gathering techniques to obtain the relevant data, the correlation among the variables involved, and a list of possible models. The next step is to select a set of methods and models that cover all kinds of knowledge and functionalities needed for the decision making process. Once the models are selected, they must be fully implemented by means of machine learning , data mining, statistical or numerical technique. After that, those models must be integrated to build the whole EDSS. The EDSS must be tested to check its performance, accuracy, usefulness and reliability, both from the user's and (AI) technology/computer scientist's point of view. If these is any wrong feature in any development stage, such as model's integration, models' implementation, selection of models, database, problem analysis etc. the developers must come back in the update th required components. When the evaluation phase is all right, the EDSS is ready to be applied to the environment. The great contribution of artificial intelligence to EDSS the integration of several methods complementing the classical statistical models/simulation , statistical analysis, linear models, etc. and numerical models (control algorithms, optimization techniques etc.) .

This cooperation makes the resulting systems more reliable and powerful in coping with real world environment systems. Date interpretation has been a principal area of research in (AI) technology since the very beginning. The most demanding problem in the environmental assessment context. Knowledge representation permits the definition of the different types of data that the existing methods adapt to the process. There is also a lot of work to clean, repair and transform the huge available quantities of raw data. Apart from this, the availability of meta-information or background knowledge is required to guide the process. Data mining is multi-disciplinary: It covers expert systems, data based technology, statistics, data visualization and unsupervised machine learning. These techniques operate at the level of data and background information, where numerous and often incompatible new commensurate pieces of information from disparate sources have to be brought together (K, Fedra, 1994).

So, it seems that in the future, (AI) technology with the increasing maturity in particular those related to knowledge and engineering, new dimensions can be assisted to users in environmental decision making are available. For example, many environmental systems are characterized both by incomplete models and by limited data. Hence, in the future, (AI) technology will be applied to predict climate change to reduce crop or fruit etc. food agriculture challenge by climate change bad influence.

● Will (AI) technology influence digital economy change to manufacturing industry ?

To understand how the manufacturing business must adapt to prosper in the technology, we need to understand how (AI) technology will change us to shape our daily habits to satisfy our expectation of products to how we shop and even the immediate of the entire process. For example, taxi services are in the crosshairs as on demand transportation services like, available of the touch of a smart phone button expand. In fact, Yellow lab, US country , san Francisco city's largest taxi company is filing for bankruptcy as the industry starts to change faster than almost anyone expected. However, at this point, its more than an app that is changing, some our taxi passengers renting taxi transportation to catch consumption behavior.

(AI) technology will influence digital economy for taxi passenger's individual customer experience, offering a growing renting taxi to catch of service and feedback opportunities when any one taxi passenger who chooses to use mobile phone app online tool to prepaid to rent any taxi more easily.

Also in the long term, (AI) technology can influence vehicles drive themselves of behavior. Already, companies like Google and GM are working on projects to bring fleets of autonomous vehicles to cities at the path of a button.

Moreover, this on-demand service model is beginning to appear across a much broader range of markets. For example , Amazon company is investing in its own fleet of trucks, planes and even drone at the same time as it pushes for same-day delivery of products. As some point, vehicles will be autonomous too. So, it seems that (AI) technique will influence any transportations choose to use digital autonomous driving technology in the future . For Amazon company case, it is not stopping of logistics. It is also aiming to automatically manage the supply of consumer home products with its recently launched Amazon replenishment service, Dash. Dash is a digital service that enables that connected derive to automatically order physical products from Amazon when supplies are running low. So, it seems (AI) technology will be applied to logistic function by digital technology method introduction in the future.

Hence autonomous vehicles will optimize industry supply chains and logistics operations through increased efficiency and flexibility. In fact, fully automated and lean supply chains will keep reduce load sizes and inventory by leveraging smart distribution technologies and smaller autonomous vehicles by machine man assistance. If Amazon continues to grow market share for online sales by reducing effort required by the consumer to place an order, when also contributing the almost immediate delivery of products to the doorstep. So, it will further fuel the trend toward on-demand derive. As Amazon company fuels the on-demand economy, consumers will expect immediacy in more parts of the digital economy. On top of speed, consumers increasing expect more personalization options.

So, (AI) technology will influence digital manufacturing, such as Amazon publishing to monitor every aspect of every process in real -time and communicating to self-optimized deep learning robotics, new methods of high volume and high customization will become possible. Then, as products merge into product platforms and even services, manufacturers have the opportunity to provide components and platforms used by smaller players. So, (AI) technology will influence manufacturing industry to choose automated SMI lines, robots installed, automation engineers.

Another future (AI) technology development can be applied to space science aspect, such as Automation engineering space in manufacturing process to achieve digital manufacturing benefits to any businesses in the future. Such as reducing cost, shortening manufacturing time, raising efficiency, shortening delivery products to client individual time. How can artificial intelligence give the need and advanced fast and evaluation methods benefits for space exploration? When US NASA (space exploration organization) achieves any space exploration missions, it will answer this question:

When is it useful to have a machine use (AI) technology to achieve a decision? After all, after millions of years of space exploration and rough 10,000 years of civilization, humans are usually quite good at making decisions in complex uncertain environments. Through, Johns Hoplains University's Applied Physical Lab. Research in (AI) technology enabled systems, which has identified three general use cases for (AI) technology to explore space mission:

First, for some tasks (AI) technology is more cost effectiveness than human. Second, (AI) technology is better suited than humans at solving some, but not all problems. Third, (AI) technology allows NASA organization's space exploration mission to develop machines that ate capable of responding faster than when a human is in the decision loop (D. Scheidt, 2012, A. Castano et. al. 2008).

So, the use of (AI) technology to enable science by observing the pace of rapidly evolving phenomena was demonstrated. It is more effectively coordinating and (AI) technology utilizing to earn economic benefits to use for space exploration mission.

However, (AI) technology also have current risk for space exploration. Today (AI) technology is immature and requires further development to reach its potential. For instance, the (AI) technology algorithms that detected the dust derive could not have identified whether the Martain weather represented a threat to the cover. Also it can not yet use instrument input to determine what, where and how to autonomously make the next space science measurement. An equally important factor limiting (AI)'s deployment is that lacks the methodology and technology to effectively test (AI) technology. So, the challenge will testing (AI) enabled system is how (AI) performance can be measured. It would be NASA organization's difficulty to find (AI) technology to develop to carry on researching any space exploration missions in the future. However, (AI) technology will be a good economic benefit choice for space exploration mission in the future.

● What is artificial intelligence potential benefits and ethical considerations?

The ability of (AI) technology systems to transform vast amounts of complex information into insight has the potential to help solve manufacturing or service challenges for human needs. However, to reap the societal benefits of (AI) systems, humans will need to trust then and make sure that which follow the same ethical principles, moral values, professional codes and social norms that we humans would follow in the same scenario, research and educational efforts as well as carefully designed regulation in order to achieve the most effort of economic benefits goals. For example, international business machines corporation (IBM) is actively engaged both competitors , in global discussions about how to make (AI) ethical and as beneficial as possible for people as social economic benefits.

(AI) is usually defined as the " capability of a computer program to perform tasks or reasoning processes " that human usually associate to intelligence in a human being. Often, it has to do with the ability to make a good decision, even when there is uncertainty, too much information to handle. As an example, play chess or complex card games of entertainment activities is believed to need some form of intelligence in a human being, as well as choosing the best medical facilities in a difficult medical case, or creating something new, such as mathematical theorem or even some form of act, or even driving automatic machine man (self driving vehicle) replacing human driving in the middle of a crowded city.

(AI) needs depends on what we consider being intelligence in the behavior of a human being act a certain point in time. If human belief about human intelligence changes and we don't believe any longer that a certain task requires intelligence, then a computer program performing that task is no longer part of (AI), it becomes just another boring computer program. So, it means that (AI) technology will replace some old computer programs, if human can invent new generation of (AI) software for any functions or activities to satisfy human needs.

As IBM, it argues intelligence. This means that we aim to build systems that enhance and scale human expertise and skills rather than replacing them. We therefore focus on practical applications of (AI) capabilities that assist people in performing well-defined tasks of needs by exploiting and wide range of (AI)-based services. We also use the term " cognitive computing" it is mean a comprehensive net of capabilities based on technology. It comprises the fields of machine learning, reasoning and decision technologies, language, speech and vision recognition and processing technologies, high performance and high efficient functions for any industries or individual consumers needs. For example, robotics, which are usually very good at doing what which are supposed to in any environment, much have public shopping center, factory etc. places which need simply services from the robot (machine man), such as cleans the floor of our houses to the robot that can work together with humans in production chains, passing through the warehouse, robots can take care of the tasks of an entire warehouse and the companion robots like Nao, Pepper, Aibo and Giraff, who can entertain use, talk to use and help elderly people to stay connected to their friends, relatives and doctors.

Google company is building automatic machine (self-driving cars) and has acquired more than 10 robotics companies. Facebook had opened whole new research facility only on (AI) research. Apply computer has developed Siri. Microsoft computer company has built a similar personalized assistant. Google has Deep mind, a UK company whose long term aim is to build general (AI) and has already great potential to win game to the world champion and IBM is investing a huge amount of resources in applying its Watson cognitive computing system to the medical domains to finance and to personalized education. In Europe, IBM is establishing new centers in Munich and Milan focused in the application of cognitive computer capabilities to the internet of things and healthcare respectively.

For example, automatic machine man (self-driving cars) are all about (AI), which used to be able to see what happens in the street (signals ,lanes, other cars, pedestrians, traffic lights, which need to able predict what other cars and pedestrians will do, and who need to be able to cope with unforeseen situations. Since, most car accidents are due to human fault, it is estimated that the adoption of self-driving cars will save about half of the lives that are usually last in car accidents.

IBM Watson company has to understand spoken language, make sense of massive amount to text , respond correctly to questions in many categories, as well as assess its own confidence in responding to such questions. In the future, (AI) technology can own question/answering capabilities that would be very useful, for example, in assisting a doctor when trying to some to the correct diagnosis for a patient and to propose the best therapy .

Intelligent machines can also rely on huge amounts of data to be used to learn how to make better decisions. This data comes from all of us over the years Facebook users have uploaded more than 250 billion pictures and every day who upload about 350 million more. Every second, we submit 40,000 google search queries. So, (AI) technology will be connected through the web from appliances to traffic lights from cars to watches. Other tasks that are very easy for humans are physical and manipulation tasks, such as walking , running, picking up an object to make its shape and location, restricted environment. But (AI) machine man technology still not able to have the general physical and manipulation capabilities even of a 6 year old.

So, it brings this question: Why do (AI) scientists need to concern ethics? Because (AI) technology is complex, information into insight has the potential to reveal long held secrets and help solve some of the world's most difficult problems. (AI) systems can potentially be used to help discover insights to treat disease, predict the whether, and manage the global economy. So, ethic issues is important to and (AI) scientists . If any one new (AI) technology research investigation could success, it will be a secret to and the (AI) scientists can not permit to their loyalty to any competitors to damage the fair (AI) technology products trading market. The country (countries) (AI) technology scientists need to concern ethic issues, who need to keep secrets for their countries economic or/and social benefits. This is moral issues to any countries/country loyalty is whose countries intangible assets. They can not sell (AI) loyalty to any their countries to assist whose economic benefits immorally.

● How can (AI) technology influence to global
health care economy development?

According to (AI) lecturer analysis, when combined key clinical health (AI) application can potentially create $150 billion in annual savings for the US healthcare economy by 2026 year. (AI) technology is re-winning modern conception of healthcare delivery. It enables machines to sense, comprehend, act and learn. So which can perform administrative and clinical healthcare functions (Accenture, 2017).

It will help health care service organizations to reduce health care cost, will improve and raise service quality and access. So, (AI) health market size will be predicted growth. (AI) applications in health care include robot-assisted surgery, virtual nursing assistant, administrative workflow assistant, fraud detection, error reduction connected machines, clinical trial participant identifier, preliminary diagnosis, automated image diagnosis and cybersecurity.

What kind of benefits (AI) technology can contribute to healthcare service? (AI) technology can deliver what many health care organizations need, such as financial and operational of labor costs, digital expectations from patient consumers how to use (AI) technology to solve interoperability challenges in any healthcare organizations. Also (AI) technology can be applied to wellness an d lifestyle management, diagnostics, delivers financially but also way of organizational and workflow improvement. So, (AI) technology will be continue to become most prevalent and adoption to healthcare organizations , which must need to enhance structure to be position to take full advantages of new (AI) technological capabilities. (AI) technology can change the nature of work and employment is rapidly changing to make the best use of both humans and (AI) talent in healthcare industry in the future. For example, (AI) technology offers a way to fill in gaps and the rising labor shortage in healthcare. According to Accenture analysis, the physicians shortage is increasing. However, (AI) technology will manufacture healthcare machine men to replace physicians in future one day(2017). Hence, (AI) technology will be invented to raise health care service staffs work efficiency and performance in any hospitals or clinics in the future.

In conclusion, (AI) technology will raise efficiency for any service or manufacturing industries in the future, although, it is possible that it will also rise low skillful workers unemployment numbers. But, the most important influence to human technological innovation will be risen and it will influence human life will be changed to be better, e.g. self drive cars, health care physician machine men, machine man cleaners etc. intelligent machine men will be manufactured to serve for our daily life. Furthermore, (AI) technological products will influence countries trading, some low technological development countries manufacturing businessmen can choose to buy any (AI) products to raise whose productivity and efficiency and reducing cost to achieve economic cost saving result. Also, GDP of trading growth income will increase to the (AI) products sale countries. Hence, it will be beneficial to economic development to both developed and developing countries both in the future as well as (AI) scientists time and money spending will be valued to continue to invest (AI) technology development for human life and economy benefits for long term.

In conclusion (AI) technology will raise macro economy growth and it can create many (AI) jobs , but it also raise the low level technological worker unemployment change. In the future, (AI) technology can be applied to digital technology to attempt to invent any new undiscovered (AI) and digital technology. So, it needs any scientists to continue to research how digital and (AI) technology can be mixed to satisfy human's future undiscovered needs.

Online technology and online book
technology influences artificial intelligence
mind development

Nowadays, online technological invention bring online book technological development. Also, artificial intelligent technological machine men had been invented to link internet to do any jobs, e.g. children can find any data from artificial intelligent machine men when the artificial intelligent machine man had been installed internet and computer function, then children can find any online books to read from the artificial intelligent machine man. Such as Japan artificial intellgent machine men had installed computer and internet function, the Japan family children can find any online books to read from the artificial intelligent machine man at Japan any families' homes conveniently. Hence, it implies that future one day, artificial intelligent machine has possible to be invented to own human's reading and/or writing abilities.

For example,online book publishing is one kind of popular internet technology. For cxample, Amazon publish is as a business model with many potential advantages, relative to a physical operation. It held out the potential of lower book inventing and distribution costs and reduced overhead. Consumers could find the books, they were looking for more easily and a variety book topic choices could be offered for sale. It can accept and fulfill orders from almost any domestic location with equal ease. And most purchasers made on its site would be exempt from sales tax. One Amazon strategy hand, it would have to make its returns and redress processes transparent and reliable, and offer other ways for clients to learn, as much about the book possible before buying. Future online book market development trend, such as Amazon, Barnes & Noble etc. online book shops.

Hence, online book store technology can be applied to artificial intelligent technology. Such as artificial intelligent machine men can apply computer technology to learn the abilities of reading and/or writing any books either on paper or on computer. Hence, it is possible that artificial intelligent machine men will have similar human's writing and/or reading books ability when they own human's mind ability. However, it bring this questions: Can artificial intelligent machine men own human's mind abilities? If they own human's mind abilities, is it mean that they can write and/or read any books? Can artificial intelligent machine men own human's mind abilities to create to write any books? Can artificial intelligent machine men own human's mind abilities to read and make any judgements or decisions more accurate than human's judgements or decisions? To answer these questions? I shall indicate that online book reading and writing technology can be applied to artificial intelligent machine men reading and writing technology. Because they are similiar computer mind technological development. So, I believe that future artificial intelligence machine men can be invented to own similar human's reading and writing's mind abilities in future one day.

I believe artificial intelligence and online technological reading abilities are very similiar. Nowadays, computer can be invented to attempt to read and write any books by human. Why can not artificial intelligent machine men replace computer to read and write any books? Artificial intelligent machine men can replace human to attempt to write or/and read books, due to artificial intelligent machine men had invented to own human mind to do some jobs and their mind had been invented to be similiar to human behavioral abilities to do these behaviors, e.g. cooking, driving, playing games, singing songs, speaking, listening, frighting etc. different human's abilities. So, it seems that artificial intelligent will be possible to be invented to own human's mind abilities to do any writing or reading behaviors or functions.

Prediction of artificial intelligence
reading and writing abilities

development

What is future trend of artificial intelligence reading and writing abilities development? To answer this question, we need to know what benefits of artificial intelligent machine men can attribute to human's needs when they can own any human's mind to read or/and write any books.

I shall indicate e-books reading and writing example, if artificial intelligent machine men can be invented to own human's mind to write and/or read e-books on computer. Then, it brings this question: Can artificial intelligent machine men assist human to learn to do judgement to solve any challenges?

I believe that when artificial intelligent machine men can be invented to own human mind to write or/and read any books, then they will own human's mind ability to make judgement to solve any challenges more accurately, even their decisions can be more accurate to compare to human's decisions. So, artificial intelligent machine mens' writing and reading ability is the main factor to cause their mind to do any judgement in order to make any decisions more accurately. Consequently, in future one day, artificial intelligent machine mens' writing and reading ability will be invented to similar human's reading and writing abilities as well as their minds can also be invented to similar human's minds as well as their judgement abilities can be invented to similar to human's judgement abilities to make any decisions more accurate.

The influences when AI is invented to own
human's mind and judgement abilities

Finally, I shall discuss what are the influences when AI is invented to own human's mind and judgement abilities in our future job market. The achievement of artificial intelligent (AI) machine men achievement requirement of owning human's mind and judgement abilities which requires extensive manual labor, and by augmenting the calling process with machine learning, the process where speed and accuracy are needed to close to human's mind and judgement abilities. Expert human race callers now have better information at artificial intelligent machine men at their fingertips faster.

Hence, if the above those requirements are achieved to satisfy artificial intelligent machine men ind and judgement abilities demand to close or exceed humans' mind and judgement abilities. Then, I believe that future human's some simple jobs must be replaced by (AI) machine men. Even, human's some professional jobs, e.g. lawyer, accountant, administator, typing etc. professional skillful jobs, which will be either replaced or will be assisted by (AI) machine men. For example, (AI) machine men learn how to type english or other language words to do typing job ; they can learn how to apply accounting knowledge to record any firm's income and expenditure record of accounting job; they can also learn how to assist architects to design any architectural building drawing plans to do architect jobs; they can learn how to analyze any court evidences to judge any criminal or civil cases and assist lawyers to give legal advices to achieve more reasonable judgement for any legal cases; they can also learn how to assist firm's managers or administrators to manage any organization teams efficiently.

Consequently, when (AI) machine men can be invented to achieve to exceed human's mind and judgement abilities level. Then, I believe that they can do instead of human' simple jobs, which can do even human's more difficult and more judgement requirement of professional skillful jobs. So, (AI) machine men must need to achieve to do any jobs, they are same, even exceed to human professionals' abilities. Then, it will cause a lot of human's jobs to be disappeared or some human's jobs will be replaced by owning judgement and mind abilities of (AI) machine men to do.

Hence, future many human's jobs will be replaced by technological labors. Employers choose to buy (AI) machine men to replace human labors. The reasons include (AI) machine men have none unhappy, angry emotion to influence their low efficiencies and low productivities. Their judgement and mind abilities can exceed human's abilities or do any jobs to compare better performance to human's abilities. Consequently, different occupation labors need to prepare to learn how to co-operate with (AI) machine men to let future employers feel (AI) machine men will be human's assistant to assist human to do jobs efficiently when human and (AI) machine men work together. It aims to avoid future employers feel (AI) machine men's judgement and mind abilities can exceed any low knowledgeable and skilful occupation labors, even high knowledge and skilful occupation labors. It means that (AI) machine men are only labors' assistant if (AI) machine mens' judgement and mind abilities are below under to human labors' judgement and mind abilities.

Consequently, to avoid (AI) machine men can replace human to do any simple or complex jobs to cause any future any occupation labors' competitors. I recommend that it is right time labors ought prepare to learn different skills. So, every individual labor does not only concentrate on one kind of skill. Because supposing one kind of the occupation labor's job duties are replaced by (AI) machine men. If the employee had owned more than one kind of occupation skill. Then, I believe that who can avoid the unemployment threat more easier than the employee only owned one kind of occupation skill, when (AI) machine men had invented to own human's mind and judgement abilities in future one day. The most important, I believe that (AI) invention will be applied to be teach how to learn human's skills and mind ability. Such as education industry, teacher won't be replaced by (AI), otherwise, (AI) will be teacher's assistant to help them to do education data gather or teaching jobs. So, teachers won't be replaced by (AI), otherwise, teachers will depend on (AI) data gather or teaching job to give them opinions how to solve student's teaching challenges as well as teachers can concentrate on researching education jobs for schools' benefits if (AI) technology can be invented to on human's mind and judgement and reading and writing abilities in the future.

Online technology and online book
technology influences artificial intelligence
mind development

Nowadays, online technological invention bring online book technological development. Also, artificial intelligent technological machine men had been invented to link internet to do any jobs, e.g. children can find any data from artificial intelligent machine men when the artificial intelligent machine man had been installed internet and computer function, then children can find any online books to read from the artificial intelligent machine man. Such as Japan artificial intelligent machine men had installed computer and internet function, the Japan family children can find any online books to read from the artificial intelligent machine man at Japan any families' homes conveniently. Hence, it implies that future one day, artificial intelligent machine has possible to be invented to own human's reading and/or writing abilities.

For example, online book publishing is one kind of popular internet technology. For example, Amazon publish is as a business model with many potential advantages, relative to a physical operation. It held out the potential of lower book inventing and distribution costs and reduced overhead. Consumers could find the books, they were looking for more easily and a variety book topic choices could be offered for sale. It can accept and fulfill orders from almost any domestic location with equal ease. And most purchasers made on its site would be exempt from sales tax. One Amazon strategy hand, it would have to make its returns and redress processes transparent and reliable, and offer other ways for clients to learn, as much about the book possible before buying. Future online book market development trend, such as Amazon, Barnes & Noble etc. online book shops.

Hence, online book store technology can be applied to artificial intelligent technology. Such as artificial intelligent machine men can apply computer technology to learn the abilities of reading and/or writing any books either on paper or on computer. Hence, it is possible that artificial intelligent machine men will have similar human's writing and/or reading books ability when they own human's mind ability. However, it bring this questions: Can artificial intelligent machine men own human's mind abilities? If they own human's mind abilities, is it mean that they can write and/or read any books? Can artificial intelligent machine men own human's mind abilities to create to write any books? Can artificial intelligent machine men own human's mind abilities to read and make any judgements or decisions more accurate than human's judgements or decisions? To answer these questions? I shall indicate that online book reading and writing technology can be applied to artificial intelligent machine men reading and writing technology. Because they are similiar computer mind technological development. So, I believe that future artificial intelligence machine men can be invented to own similar human's reading and writing's mind abilities in future one day.

I believe artificial intelligence and online technological reading abilities are very similiar. Nowadays, computer can be invented to attempt to read and write any books by human. Why can not artificial intelligent machine men replace computer to read and write any books? Artificial intelligent machine men can replace human to attempt to write or/and read books, due to artificial intelligent machine men had invented to own human mind to do some jobs and their mind had been invented to be similar to human behavioral abilities to do these behaviors, e.g. cooking, driving, playing games, singing songs, speaking, listening, frighting etc. different human's abilities. So, it seems that artificial intelligent will be possible to be invented to own human's mind abilities to do any writing or reading behaviors or functions.

Prediction of artificial intelligence
    reading and writing abilities

development

What is future trend of artificial intelligence reading and writing abilities development? To answer this question, we need to know what benefits of artificial intelligent machine men can attribute to human's needs when they can own any human's mind to read or/and write any books.

I shall indicate e-books reading and writing example, if artificial intelligent machine men can be invented to own human's mind to write and/or read e-books on computer. Then, it brings this question: Can artificial intelligent machine men assist human to learn to do judgement to solve any challenges?

I believe that when artificial intelligent machine men can be invented to own human mind to write or/and read any books, then they will own human's mind ability to make judgement to solve any challenges more accurately, even their decisions can be more accurate to compare to human's decisions. So, artificial intelligent machine mens' writing and reading ability is the main factor to cause their mind to do any judgement in order to make any decisions more accurately. Consequently, in future one day, artificial intelligent machine mens' writing and reading ability will be invented to similar human's reading and writing abilities as well as their minds can also be invented to similar human's minds as well as their judgement abilities can be invented to similar to human's judgement abilities to make any decisions more accurate.

The influences when AI is invented to own
human's mind and judgement abilities

Finally, I shall discuss what are the influences when AI is invented to own human's mind and judgement abilities in our future job market. The achievement of artificial intelligent (AI) machine men achievement requirement of owning human's mind and judgement abilities which requires extensive manual labor, and by augmenting the calling process with machine learning, the process where speed and accuracy are needed to close to human's mind and judgement abilities. Expert human race callers now have better information at artificial intelligent machine men at their fingertips faster.

Hence, if the above those requirements are achieved to satisfy artificial intelligent machine men ind and judgement abilities demand to close or exceed humans' mind and judgement abilities. Then, I believe that future human's some simple jobs must be replaced by (AI) machine men. Even, human's some professonal jobs, e.g. lawyer, accountant, administator, typing etc. professional skillful jobs, which will be either replaced or will be assisted by (AI) machine men. For example, (AI) machine men learn how to type english or other language words to do typing job ; they can learn how to apply accounting knowledge to record any firm's income and expenditure record of accounting job; they can also learn how to assist architects to design any architectural building drawing plans to do architect jobs; they can learn how to analyze any court evidences to judge any criminal or civil cases and assist lawyers to give legal advices to achieve more reasonable judgement for any legal cases; they can also learn how to assist firm's managers or administrators to manage any organization teams efficiently.

Consequently, when (AI) machine men can be invented to achieve to exceed human's mind and judgement abilities level. Then, I believe that they can do instead of human' simple jobs, which can do even human's more difficult and more judgement requirement of professional skillful jobs. So, (AI) machine men must need to achieve to do any jobs, they are same, even exceed to human professionals' abilities. Then, it will cause a lot of human's jobs to be disappeared or some human's jobs will be replaced by owning judgement and mind abilities of (AI) machine men to do.

Hence, future many human's jobs will be replaced by technological labors. Employers choose to buy (AI) machine men to replace human labors. The reasons include (AI) machine men have none unhappy, angry emotin to influence their low efficiencies and low productivities. Their judgement and mind abilities can exceed human's abilities or do any jobs to compare better performance to human's abilities. Consequently, different occupation labors need to prepare to learn how to co-operate with (AI) machine men to let future employers feel (AI) machine men will be human's assistant to assist human to do jobs efficiently when human and (AI) machine men work together. It aims to avoid future employers feel (AI) machine men's judgement and mind abilities can exceed any low knowledgeable and skilful occupation labors, even high knowledge and skilful occupation labors. It means that (AI) machine men are only labors' assistant if (AI) machine mens' judgement and mind abilities are below under to human labors' judgement and mind abilities.

Consequently, to avoid (AI) machine men can replace human to do any simple or complex jobs to cause any future any occupation labors' competitors. I recommend that it is right time labors ought prepare to learn different skills. So, every individual labor does not only concentrate on one kind of skill. Because supposing one kind of the occupation labor's job duties are replaced by (AI) machine men. If the employee had owned more than one kind of occupation skill. Then, I believe that who can avoid the unemployment threat more easier than the employee only owned one kind of occupation skill, when (AI) machine men had invented to own human's mind and judgement abilities in future one day.

How does (AI) robots' brain invention influence our lives

(AI) research modeling the human brain has developed important technologies, and has overcome significant barriers. How will (AI) affect humanity in the near future? How will (AI) change our lives and our societies? Is the evolution of (AI) to humanity, or it represent a threat?

On white collar workers (AI) job replacement aspect, University of Tokyo, Institute of informatics, lecturers who had attempted to do experiments to take (AI) exams over a two year period. The (AI) achieved standard scores of around 50 in each subject, exceeding the norms for humans attempting the tests. The (AI)'s results in subjects emphasizing memorization, such as world history and Japanese history subjects were comparatively high, and the results of the study suggested that an appropriate selection of subjects would give at an 80% chance of passing the entrance exams of 80% of Japan's private universities.

So, if (AI) is applies to human white collar workers' job duties aspect, at this level, if white collar workers were replaced by (AI) in the future, around 30% of current staff would be replaced. Whatever, the outcome, large companies will be represented with two choices. One choice will be to protect their employees, but as a result lose their international competitiveness. The latter choice will enable them to reduce the cost of general duties, financial management procedures, accounting etc. general administrative job duties of cost in offices. Hence, it seems that (AI) will be possible to be invented to own human's brain ability to do some mind jobs in future on day.

However, the method called " deep learning" must be developed to cause (AI) to match human's brain ability as well as these were dramatic advances in technologies, such as image recognition and voice recognition, which form the foundation for (AI). Nowadays, this new method called" deep learning" does not reach the matured and stagnated stage. It needs to wait human to continue to invent to let (AI) to match human brain to achieve 100% owning human's mind ability. Nowadays, (AI) industry product include cleaning robots, smart TVs and future (AI) product development market. It will include self-driving vehicles, drones, and nursing robots.

On (AI) weapons applied aspect, if (AI) can be invented to own human's brain judgement and analytical abilities. Then, it is possible that it can be applied to attack enemy to cause war effect. For example, if weapons such as missiles were equipped with (AI) in the future, they would become able to decide on their own targets. Hence, human needs to apply restrictions when necessary.

On (AI) applied to analyzing information collected technological aspect, nowadays, every one will use wearable terminals to connect to the internet to obtain various types of information as well as computers will collect and analyze information on people. Our lives will probably be more reliant on these internet technologies than they are on smartphones today. When, (AI) can match human brain to own mind ability.

Then, (AI) can be applied to do any analyzing information and collection job duties aspect to raise large information restoring and remembering efficiency. For example, (AI) will be generally used and will be extremely useful in analyzing the information collected from wearable devices and stored in the cloud. (AI) will enable wearable devices to be of real assistance in our lives offering their users more intelligent support.

Rather than allowing (AI) to develop on serves, as something separate from humanity. It will be more meaningful to encourage its development via wearable devices, situating it under the control of human intelligence. The intelligence of (AI) will increase rapidly in the future. If this increase in (AI) occurs under human control, enabling humans to increase their own abilities, then surely it will be possible for us to put up a degree of resistance to the opposite scenario, the domination of (AI) over humanity. Hence, if (AI) can be invented to remember and store and make analytical judgement to collect any information from internet. Then, it will bring the effect, such as large international organizations' (AI) internet storage robots can bear in mind factors, such as competitors' privacy or business secret information, such as the loss equality between people and threats to privacy that will be stolen form the owning (AI) storing internet information remembering robots.

Consequently, what is the effect of successful invention of (AI) matching human brain's mind ability? (AI) present computers are adequately able to reproduce the emotional, conceptual and intuitive abilities of humans. Because of this, it is important that we should envision potential future problems that may manifest when we consider how to employ wearable devices. It will be essential to enhance our technologies in order to ensure that we can use (AI) under human control.

However, when a goal has been set. (AI) will implement an appropriate means for its realization. (AI) will be need as a tool by human society. If the capacities of analytical and judgement mind abilities of (AI) brain exceed those of human brain, it is difficult to imagine the type of technological, then singularity is represented by the creation of an (AI) by another (AI). It is important that we rapidly and accurately predict these developments, when image recognition and other individual technologies are functioning at a high level. There will be a considerable matter in different sectors of (AI) industry development.

Today, however, machines have become able to decide for themselves what they will learn, making it difficult to copy human's mind ability. What we must consider when machines exceed humans and (AI) surpasses human capabilities. May technologies exceed human capabilities, cars are faster than humans, planes are able to fly. Consequently, it brings a question that human needs to consider: When does (AI) brain technology be invented to reach the most reasonable stage to be accepted or stopped by humanity?

What is artificial intelligence
human brain invention?

A machine is likely to achieve the ability of a human brain. Does it a scientific story? Some scientists has predicted that a US$1,000 personal computer will match the computing speed and capacity of the human brain by around the year 2020 year. With human reverse engineering, human should have the software insights before 2030 year. it is possible that of machine intelligence and exotic new technology for faster and more powerful computational machines from cellular automata and DNA playing cheese game competition case example, it proves that (AI) had been invented to own human's analytical and judgement ability to exceed the best cheese game human player's brain analytical and judgement ability. Then, it seems that (AI) will have possible to be built machine brains to achieve the exceed level of human brain's analytical and judgement ability in the future one day.

Supposing we scan someone's brain and restate the resulting " mind file" into suitable computing medium. Will the entity that emerges from such an operation be conscious? How have advances in electronic communications changes power relationship? For electronic book publishing case example, a book that looks at the principles companies must adopt to meet the needs and desires of this new kind of client. So, such as paper book can be changed to electronic book for human to read. Why can't human brain be changed to (AI) machine brain to do human's analytical mind and behavioral mind of activities to replace to do any human's daily analytical and behavioral mind activities?

Over the next few decades, machine achieve super intelligence, human will encounter a dramatic phase. Will it be a "WALL" a barrier as conceptually the event of a black hole in space. Such as (AI) brain invention case, an " AI singularity" ruled super-intelligence AIs, or a gentler " surge" into a post human era of agelessness and super-intelligence brain. Will future technology, such as bio-engineered pathogens, self replicating nan robots, and super smart robots run and accelerate out of control, perhaps threatening the human race?

If one day, (AI) brain is invented to achieve agelessness possibility. It means human's brain will be old to lose mind and analytical ability when human's age is increasing. Otherwise, (AI) machine brain age won't lose mind and analytical ability, due to (AI) machine is no age increasing possibility. It is a machine brain. If (AI) machine brain can be built successfully. Scientists need to consider technological ethic matter, such as the challenge of guiding nanotechnology in a constructive direction, advances in nanotechnology and related advanced technologies can not be inevitable, any broad attempt to relinquish nanotechnology would interfere with the benefits. When actually making the dangers worse.

Keeping in mind that intelligence machines are already making their way into our blood stream. There are dozens of projects underway to create blood-stream based " biological micro electronic- system" (bio MES) with a wide range of diagnostic and therapeutic applications BioMEMS devices are being designed to intelligently pathogens and deliver medications in very precise ways. For example, a researcher at the University of Illinois at Chicago has created a ting capsule with pores measuring only seven nanometers. The pores let insulin out in a controlled manner, but prevent antibodies from invading the pancreatic Islet cells inside the capsule. These nano- engineered devices have cured rated with type I diabetes, and there is no reason that the same methodology would fail to work in humans. Similar systems could precisely deliver dopamine to the brain patients, provide blood-clotting factors for patients with hemophilia and deliver cancer drugs directly to tumor sites. A new design provides up to 20 substance-containing reservoirs that can release their cargo at programmed times and locations in the body.

Another brain health technological related invention case, such as Kensall Wise, a professor of electrical engineering at the University of Michigan, who has developed a tiny neural probe that can provide precise monitoring of the electrical activity of patients with neural disease. Future designs are expected to also deliver drugs to precise locations in the brain. Also, kazushi Ishiyama at Tohoku University in Japan has developed micro machines that use microscopic-cancer tumors.

A particularly innovative micro machine developed by Sandia National labs has actual micro teach with a jaw that opens and closes to trap individual cells and then implant them with substances, such as DNA, proteins or drugs. There are already at least four major scientific conferences on bio MES and other approaches to developing micro-and nano-scale machines to go into the body and bloodstream. All these inventions are related to how to apply machines to copy human's brain knowledge in order to achieve to do any human's brain functions.

Finally, for Freitas envisions micron-sized artificial platelets invention case example, who could achieve hemostasis (bleeding control) up to 1,000 times faster than biological platelets. Freitas describes nano-robotic microbivores (white blood cell replacement) that will download software to destroy specific infections hundreds of time faster than antibiotics, and that will be effective against all bacterial, and fungal infections with no limitations of drug resistance.

Consequently, such as above machine health scientific invention cases, there were many scientists had invented any health machines to apply drugs to transfer to human's brain to attempt to reduce human's disease causing risks, such as reducing cancer cell increasing number. Why it is no possible that scientists can attempt to invent (AI) brain which can own human's mind ability to judge or analyze any matters to give opinions in order to exceed human's judgement and analytical ability.

How can artificial intelligent brain

satisfy to human beneficial and
natural needs?

Nowadays, new scientists' most familiar form of this vision in our times is genetic engineering. Specifically, the prospect of designing better human beings by improving their biological systems of a small, serious and accomplished group of tailors in the field of artificial intelligence and robotics. Their goal is a simply new age of post-biological life, a world of intelligence without bodies, immortal identity without the limitations of disease, death and unfulfilled desire. If human can understand why this fate is presented as both necessary and desirable, human might understand modern science can help us to enter the good life and good society stage when (AI) brain is invented by scientists successfully in our future life.

How can (AI) beneficial brain satisfy to human natural need? For relatively recent example, similarly as a long term trend beginning with the first mechanical calculators, the evaluation of computing capacity increases in speed over time and decrease in cost. From biological evolution has been invented to influence human brain, an electronic chemical machine with a great, but finite number of computer neuron connections, the product of which we call mind or consciousness. As an electro-chemical machine, the brain obeys the laws of physics, all of its functions can be understood and duplicated. And since computers already operate at far faster speeds then the brain, they soon will rival or surpass the brain in their capacity to store and process information. When happens, the computer will at the vary least, be capable of responding to stimuli in ways that are indistinguishable for human responses. At that point, we would be justified in calling the machine intelligent, we would have the same evidence to call it conscious that human now have when giving such a label to any consciousness other than our own.

At the same time, the study of human brain will allow us to duplicate its functions in machine circuitry. Advances in brain imaging will allow us to " map out" brain functions, allowing individual minds to be duplicated in some combination of hardware and software. The result, will be a world that is remade and reconstructed at the atomic level through nanotechnology, a world whose organization will be shaped by an intelligence that surpasses all human comprehension.

Whether or not today's humans are willing or able to " download" their brains into machines, there will come a time when all human beings will be intelligent machines in the future. Computer hardware will continue to get faster, cheaper and more powerful computer software will increase in sophistication. Brain research will continue to explore the " mechanics" of consciousness. Nanotechnology will continue to develop.

There are powerful incentives, commercial, military, medical and intellectual that will drive many of the advances that the extinctions desire if for very different reasons. Much of the work in artificial intelligence and robotics is open to the same defense that is made on behalf of biotechnology: If we don't do it, they will and why suffer or be unhappy when some new agent or invention is available that will solve the problem.

Finally, we already accept significant artificial argumentation and replacement of natural body parts when those parts are missing or defective. Over time indistinguishable from or " superior" to their biological counterparts as they employ increasing computer processing power. There are powerful incentives, commercial, military, medical and intellectual that will drive many of the advances that the extinctions desire, if for very different reasons. Much of the work in artificial intelligence and robotics is open to the same defense that is made on behalf of biotechnological if we don't do it, they will and why suffer or be unhappy, when some new agent or invention is available. That will cure the problem.

Finally, we already accept significant artificial augmentation and replacement of natural body parts when those parts are missing or deductive. Over time, such replacements are only likely to get more useful and perhaps eventually indistinguishable from or superior to their biological counterparts, as they employ increasing computer processing power. Nor is there an obvious distinction between using manufactured chemicals to fight disease and using " smart" nanotcchnology. The extinction project is begun by offering new routes to fulfilling old promises about doing good for human beings. But, it doesn't necessary end.

In connection with machine intelligence, it does not seem very promising to try to limit the power or ability of computers. The danger (or promise) that computers might develop characteristics that lead some people to call them conscious and that this age of intelligent machines would mean our extinction seems remote when compared with their practical benefits. We already rely so heavily on computers that the incentives to make them easier to use and more powerful are very great. Computers already do a great many things better than we can, and there seems to be no natural place to enforce a stopping point to further abilities. Certainly mechanistic and reductionist assumptions about society, ethics and psychology the notion that we are atoms or animals, driven by chance or instinct, run deep in the present world.

Artificial intelligent brain future
innovation and attribution (AI) brain invention successful factors

(AI) brain will bring what attribution to influence human's positive impact. How (AI) brain will be invented to apply to any businesses' needs. By how much the (AI) brains might exceed us remain unknown, but it could potentially be by a very significant degree. Future (AI) brain invention will have noted similar growth in everything from hard-drive storage density to the price and speed of DNA sequencing. On key feature of this technological growth that has not been adequately measured in the degree to which technology is becoming more intelligent.

(AI) brain test experiment
When there is an intuitive sense that (AI) programs/brains today are more capable than those of age, and that those were considerably " smarter" than the serial instructions that passed through the first supercomputers. Scientists will carry on testing (AI) to do any experiments to improve (AI) brain development. They will assess the progress of artificial (general) intelligence, but the need for intelligence tests and tests for other cognitive abilities will be tested in the forthcoming decades for bots, robots, avators, " animats" etc. and any collective system of these and biological systems (humans and non-human animals).
When (AI) brain technology is invented successfully, the idea of a super intelligent computer means invention successfully also. Whether super intelligent computer or (AI) brain can be invented successfully. Similarly, many think that the future beyond the technological singularity is unknowable or even unimaginable. Since its conception, the idea has been examined and explored by technologists. (AI) brain invention will be mean human-equivalent (AI) vs human-level(AI). Many machine intelligence tests is the exclusive focus on identifying systems that achieve human equivalence. Considering our experience with studying non-human animal intelligence.
The need to distinguish human equivalent (AI) from human level (AI) seems critical. Perfect human intelligence , such as (AI) brain invention is likely to be extremely difficult to achieve. Potentially every aspect of the biological processes involved would need to be translated with very high

fidelity.

Human level (AI), such as (AI) brain invention is another matter. Achieving capabilities that are equivalent to those of the human mind could be very feasible if it is not limited to perfectly processes involves. For instance, some pattern recognition algorithms are already superior to human abilities. This machine ability is not achieved by duplicating the processes our brains use, through some of the methods have been inspired by them. If researchers had been limited to replicating the brain's mind processes, We would still be waiting for the development of a (AI) brain machine equivalent.

Tests that would seen to have a reasonable chance of successfully testing non-human intelligence are those that use mathematics to define the value of a given challenges for (AI) brain invention. Such as Capability-test has some potential to generate meaningful data about (AI) machine intelligence, e.g. others in complexity theory test, it presents a series of abduction and prediction problems, similar to those in standard IQ tests to (AI) brain. Hence, IQ test and C-test will be potential tests to test (AI) brain ability.

(AI) brain test goals include that to identify a range of possible types of mind such as: super-fast human mind, mind with operational access to its source code, any mind capable of general intelligence and self awareness, general intelligence without self-awareness, self-awareness without general intelligence, super-logic , machine without emotion, mind capable of imaging greater mind or creating greater mind to compare human's brain abilities. Thus, any IQ or Capability test experiments aim to evaluate whether (AI) brain mind ability which can exceed to human brain mind ability. Thus, if future one day, scientists could prove (AI) brain mind ability can exceed to human brain mind ability. Then, they believe (AI) brain invention has ensured to achieve success.

(AI) brain invention successful factors

Hence, scientists want to invent (AI) brain successfully. They need to solve these challenges. Such as How robots and computers have progressively supplemented humans, initially only in relatively simple computational and manipulation tasks, but more recently in higher cognitive tasks that used to be the pre negative of the human brain, including language, mathematics, probabilistic reasoning and decision making.

An important challenge is how to enhance the productive interactions between humans and artificial intelligence? The important challenge include these major successful factors to invent (AI) brain technology, such as:

What is the state of the art in (AI) software and machine learning?

Can all aspects of brain function be manufactured by artificial system?

What is the proper form of mathematics that may capture the operation of minds and brains?

How to make (AI) brain feels consciousness?

What would it be taken for a machine to pose a sense of art in (AI) brain software and (AI) brain machine learning by artificial system?

Will machines soon surpass us in all domains of human competence?

What is the proper form of mathematic that may capture the operation of minds and brains?

What is consciousness to (AI) brain invention?

Could a machine be endowed with an artificial consciousness?

What would it take for a machine to posses a sense of self?

Will intelligence machines soon pose a danger to humanity of (AI) brain is invented to reach the mature stage successfully?

Is it possible to design and construct an intelligent robot with an artificial brain sense of ethics?

How can we enhance the humanitarian uses of artificial intelligence brain and owning mind ability of robotics, in particular in the field of education, health and emergencies?

Consequently, above all these challenges, I recommend scientists need to solve to achieve (AI) brain invention in order to reach (AI) brain invention mature stage easily.

Artificial intelligence brain invention opportunities
and challenges

If scientists focus wrong direction to invent (AI) brain, it will bring wrong marketing development to attribute any benefits to human. So, they need to reduce a mismatch of timescales between the pace of commercial innovation and (AI) brain invention process, reduce an underappreciation of the fundamental unpredictability of (AI) brain autonomous systems and reduce a lack of a university agreed upon conceptual framework for any (AI) brain invention and reduce a disconnect between the (AI) brain design of any kind of (AI) autonomous robots. Thus, scientists need examine these gapes, provide a roadmap of opportunities and challenges and identify areas of any (AI) beneficial functions to be attributed to human to use for our daily life needs.

Future (AI) brain market opportunities, it is rapidly growing innovations in digital -electronic and information technology had development of new intelligence, surveillance, and reconnaissance platform and battle management capabilities, precision-strike weapons, stealth aircraft, smart weapons and sensors and tactical exploitation of space (e.g. GPS).

Any one of these aspects will be (AI) brain future marketing development opportunities. Future (AI) brain invention of so called " narrow AI", (i.e. non-sentient artificial intelligence, whose problem-solving capability is confined to one narrow task. For example, (AI) brain needs to find the best method to win any one of chess player in any both human and (AI) robot chess playing game.

In smart weapon strategy industry, (AI) brain is needed to design how to analyze or mind to protect whose country to avoid enemy attack in any war by the best weapon protection strategy. In education industry, (AI) brain is needed to design how to analyze or mind how to assist teachers to educate whose students by the best education method. In aircraft industry, how to design (AI) brain to analyze or mind to assist pilot to make the most correct flying direction judgement to fly in the most safe way. In space exploitation industry, (AI) brain is needed to design how to find undiscovered natural resources to supply to human to use in anywhere space.

Thus, future (AI) brain invention needs have these features/characteristics to be designed. They include: (AI) learned on its own, where to find the information it needs to accomplish a specific task, (AI) can predict the immediate future from studying any matter, (AI) automatically needs to be inferred the rules that govern the behavior of individual robots within a robotic swarm simply by watching, (AI) needs to be learned how to navigation the acquired memories and experiences, much like a human brain, (AI) speech recognition needs to be reach human parity in conversational speech, (AI) communication system needs to be invented its own encryption scheme, without being taught specific cryptographic algorithms (and without revealing to researchers how its method works, (AI) translation algorithm needs to be invented to remember fluent language to more effectively translate between any two languages (without being taught to do so by humans), (AI) brain system interacted with its environment (via virtual environment) to learn and solve problems in the same ways that a human child can do, (AI) based medical diagnosis system needs to be achieved 99% percent accuracy in any medical reviewing researches (at a rate minimum 30 times faster than humans), (AI) poker playing program brain development needs to be defeated some of the world's best human poker players during a minimum three-week-long tour, (AI) brain development needs to be effectively " read minds" of human test subjects looking at pictures of faces, via functional magnetic reasonable images of brain activity.

Consequently, (AI) future brain development needs to follow above directions to be invented to attribute to human's satisfactory needs.

(AI) brain legal remembering attribution

Future, (AI) robots can assist lawyers to deal any legal cases more efficient. If human understands that smart (AI) technology is not to replace human lawyers, but to make a better lawyer that forces who to use, emotional intelligence, and capital on lawyers' mind, then human lawyer have made the first step in future, proofing whose legal service business from (AI) legal robots' assistance. (AI) promises to be a real advantage for today and tomorrow' lawyers having to deal with the rate of legislative evolution and technological change.

In future, for the better with regard to technology in any law firms. According to the ALM 205 law tech. survey 95% of firm leaders and technologist respondents agreed with recent decisions by management regarding the firm's technology in legal service profession.

How can (AI) robots be applied to legal service industry by legal service firms? The (AI) reality in the legal world, it includes in relation to the four key elements of legal service provision, such as commodity, research, reasoning and judgement, exist and can support or replace certain aspects of every lawyer individual jobs both fee earning processes and business processes can be supported and/or replaced by expert systems, cognitive computing, robotics automated systems, (AI) and the machine learning, clients demand and expect more speedy, accurate, expert, creative, intuitive and accessible legal advice, it can assist young lawyers to innovate / tech. focused firm, reducing pressure both from within law firms (or in house teams) and from clients to respond to the demand for client-designed service from (AI) robots assistance, technology related projects that are both user and client -centric need to be implemented successfully.

Due to the deployment of (AI) robots in the legal ecosystem where lawyers, firms, general counsel and clients are beginning to believe (AI) capability technologies, (AI) robots can assist lawyers to increasingly become more productive, efficient, accurate, better quality, less labor intensive and time intensive and the role of the lawyer is gradually changing. If we break down a lawyers' tasks in a legal project from beginning that can handle the majority of these four tasks far more quickly and accurately than human lawyer. (AI) robots can handle legal task in the four aspects as below:

First in legal aspect, it can be used for deep research and processing, such as extracting specific pieces of information from land registry documents, (AI) technology is placed top of a document set including client guidelines and similar forms. It searches through documents and extracts key data points to provide a report of data for improved coordination with clients.

Second on managed services technology aspect, (AI) platform which could have a huge advantage for general counsel and law departments in corporations and for clients of all company sizes.

Third on reading aspect, (AI) brain program that reads and analyzes , e.g. clauses in loan agreements. Its program helps its lawyers through transactions and points. Then toward the correct precedents of each stage of a process.

Fourth, on academics aspect, (AI) robots can assess the merits of personal injury cases. It can automatically review high volumes of contract documents to identify provisions that could potentially be impacted by contract law regulation.

Thus, all of these systems can handle large quantities of structured and unstructured data, and assist with the process management, research, and reasoning elements related to legal issues. Thus, in future, (AI) brain development can be invented to apply to knowledge research job, such as legal industry.

Artificial legal intelligence presents a thought-provoking approach to both computational models of legal reasoning and the use of evolutionary thinking about the law. The visions of computerized artificial legal intelligence , a vision of developments in both technology and legal history. A number of creative research projects have applied artificial intelligence techniques to the domain of legal reasoning.

Consequently, (AI) brain for legal industry marketing development ought concentrate on those three fields of artificial intelligence at most relevant to work in the legal areas in order to achieve the excellent attribution. Such as case-based reasoning, expert systems and neural networks. Artificial intelligence program, such as the legal reasoning programs. Thus, (AI) brain invention needs to own these human legal concept knowledge in order to achieve (AI) legal brain program development successfully.

Whether human mind can create to (AI) brain mind

How can (AI) scientists build a machine that think? In fact, there was general agreement that minds can be existence on non-biological substrates and that algorithms are of central importance to the existence of minds. However, there are much debate about the raw hardware power present in organic brains, such as (AI) brain.

I think (AI) scientists need to invent powerful hand ware, e.g. commercial digital signal processing might be, giving an appearance even to digital operations, but nothing would ever make up the intellectual runaway that is the essence of the singularity.

I also think raw hardware power is not be able to organize the parts to behave in a super-human way as well as I also think powerful software complexity is the main factor to solve (AI) brain mind invention challenge. Hence, future super-human (AI) brain invention will ought consider how to invent superhuman software more than hardware, because software can store any memory, i.e. human mind. When, (AI) brain invention which can achieve to own human mind ability. Then, (AI) scientists need to consider ethic matter: Does the future of (AI) pose an existential threat to humanity? How do we present learning algorithms from morally objectionable biases? Should autonomous (AI) be used to kill in warfare? How should (AI) systems be in our social relations? Is it permissible to fall in love with an (AI) system? What sort of ethical rules should (AI) like a self-driving car use? Can (AI) systems suffer moral harms? All those ethic matters. I think (AI) scientists need to consider after (AI) brain owns human's mind ability because it is possible that (AI) robots will harm human if they are educated to do any wrong or illegal or immoral mind ability. Thus, (AI) scientists need to consider (AI) robot's moral mind and judgement behavior.

Brain-inspired intelligent robotics

How to solve fundamental problems in the areas of brain sciences and brain-inspired intelligence technology? One way scientists seek to accomplished the mission is to develop brain-inspired hardware including intelligent devices, chips, robotic systems and brain inspired computing systems. (AI) scientists ultimate goal, but since the robot's memory and learning after the human brain, there is still much to learn about neurobiology before that goal is attached.

In (AI) brain university research aspect, two schools of thought have emerged in robotics: bio-logically robots that include a body, sensor and actuators, and brain-inspired computing robot.

Robots have found increasing applications in industry, service and medicine, due in large part to advances achieved in robotics research over the past decades, such as the ability to accomplish complex manipulations that are essential for automated product assembly. However, robots still have these weaknesses which need to be solved if (AI) scientists expect (AI) brain invention can be success. Robots still lack truly flexible movement, have limited intellectual perception and control, and not yet able to carry out natural interactions with human. These deficits are especially critical in service robots. A critical concern of government, academia, and industry is how to advance research and development for the key technologies that can bring about the next generation of robots. Developing robots with more flexible manipulation, improved learning ability and increased intellectual perception will achieve the main goals for (AI) scientists' solutions.

Consequently future (AI) brain -inspired intelligent robotic invention needs have these competitive or attractive abilities or strengths to compare computer storage ability. Such as (AI) brain needs have perception to exceed computation ability, adaptation exceeds computing speed, flexibility exceeds, computer memory access speed, cognition exceeds computer memory lifetime, learning exceeds computer memory capacity and innovation exceeds computer memory storage abilities. Hence, (AI) brain innovation must need to exceed general computer storage, lifetime, capacity, learning abilities if (AI) scientists expect (AI) robots will be popular to be applied by any service, manufacturing, education industries.

How can web intelligence need (AI) brain informatics?

Brain informatics (BI) invention will be a new inter-disciplinary field that systematically studies the mechanisms of human information processing from both the macro and micro view points by combining experimental cognitive neuroscience with advanced information technology. (BI) studies human brain from the viewpoint of informatics (i.e. human brain is an information processing system) and uses informatics (i.e. WI centric information technology) to support brain science study. It seems that (AI) scientists need to further understand how human intelligence and brain sciences development through brain sciences fosters innovative web intelligence research and development because innovative web intelligence research will have ability to assist (AI) scientists to research how to invent (AI) brain robotic technology more easily. The synergy between (web informatics) (WI) and (brain informatics) (BI) advances our ways of analyzing and understanding of data, knowledge, intelligence, and wisdom, as well as their interrelationship, organizations and creation processes. Web intelligence is becoming a central field that information technologies and artificial intelligence to achieve human level web intelligence.

(WI) may be viewed as applying results from existing disciplines , e.g. artificial intelligence (AI) and information technology (IT) to a totally new domain the world wide web. (WI) may be considered as an entrancement or an extension of (AI) and (IT), (WI) introduces new problems and challenges to the established disciplines.

Thus, developing human-level web intelligence will be seem to develop brain level artificial intelligent technology. Because brain informatics (BI) is an interdisciplinary field to systematically investigate human information processing mechanisms from both macro and micro points of view by cooperatively using experimental, computational, cognitive neuroscience, and advanced (WI), centric information technology. It attempts to understand human intelligence in depth, towards a holistic view at a long term, global vision to understand the principles and mechanisms of human information processing system (HIPS).

So, I recommend (AI) scientists needs to research how to invent brain informatics technology and web informatics technology to achieve how to apply (AI) brain to analyze and judge or mind web informatics ability to compete internet (web site) communication tool product industry. Hence, if (AI) brain can be applied to web informatics industry will increase attraction to an internet user market in the future.

Why model of sustainable development environment industry be (AI) robot attractive market

In the world scientific literature various conceptions of sustainable development are found, however, their basis are formed by three dimensions: environmental, economic and social development. The biggest attention is concerned to the environmental dimension which forms the basis of existence of social environment and economy.

The reason is emphasized that each person must preserve and manage natural resources as the basic of economic and social development. However, in the sustainable development strategy, the sustainable development is understood as among environment protection, economic and social society that form the basis to achieve the universal welfare for present and future generations, it is difficult to let human to adapt to live in pollution environment. Thus, environment pollution will be human's concerning issue. It implies how to protect environment pollution which will be human's future challenge. If (AI) brain invention can be applied to how to solve environment pollution challenge. It is very attractive attribution to human (AI) brain environment protection function can be invented to include such as : how to apply (AI) brain to do mind to give opinions or do any environment protection behaviors/activities to assist human how to effective use of natural resources, how to effective use of universal economic society's welfare, do strong social guarantees during the period of strategy's implementation (until 2020 year to achieve global (AI) robots environment protection mission from (AI) robots' behavioral assistance. Because environment pollution will influence world's climate to be worse to cause our food or vegetable can not grow up easily. Then, human will face food / vegetable shortage challenge. If (AI) brain invention can be applied to solve environment pollution aspects, then human will avoid food / vegetable shortage crisis.

How to invent (AI) brain's environment protection ability?

All similar problems significantly promotes scientists to research for new technologies of data extraction progressive software of data extraction operating on the basis of artificial neural networks allow finding the relations among various types of data in the huge data. Due to the data extraction technologies, it is possible to prove empiric observations to group, to process and model big amounts of data, distinguishing unknown schemes in data and using them in future activities (Rotman M. J., 1995).
It seems scientists believe gather every country's climate environment data to predict environment climate change will have chance to avoid environment pollution challenge. In addition, the traditional statistic methods applied to process digital data of the indicators of sustainable development environmental dimension are not able to analyze the present situation of environmental dimension in the context of sustainable development to present possible reasons of change of environmental dimension problems to generate future forecasts characterized by a high level of accuracy.
Consequently, if (AI) scientists can apply (WI) wcb informatics technology

and (BI) brain informatics technology both to apply to (AI) brain technology to gather global climate environment daily change data. Then, I believe (AI) brain invention can assist human to solve future environment pollution challenge Then, it is possible that (AI) brain can attribute to environment protection or how to give opinions or choose to do any environment pollution activities to avoid global warming crisis.

(AI) Asia market

When (AI) brain is successful invention, I think Asia will be a new market to need (AI) brain attribution for Asia consumer needs. I believe Asia people expect (AI) can attribute to let them to use. The (AI) ability includes the ability of machines and systems to acquire and apply knowledge, and to carry out intelligent behavior.

This includes a variety of cognitive tasks (e.g. sensing, processing, oral language, reasoning, learning, making decision) and demonstrating an ability to move and manipulate objects according). So, future Asia people (AI) consumers expect (AI) robots can attribute to whose society's needs, such as: how to apply intelligent systems to use a combination of big data analytics, cloud computing, machine-to-machine communication and the internet of things (IOT) to operate and learn. Asia people expect (AI) robots can give beneficial attribution to them, e.g. talking or playing a game for Asia young entertainment market; (AI) robots need to reflected by physical substance (such as any talking or playing game robot player).

In this sense, (AI) is like a human brain. For Asia (AI) robot service industry need, Asia (AI) robot clients expect to use soft robotics (robotic process automation) can be used to meet Asia (AI) robot service consumers' expectation of automation repetitive tasks and common processor needs, such as client servicing and sales without the need to transform existing IT system maps (e.g. (AI) salespeople robots, or (AI) service robotic).

In (AI) office task aspect, Asia office consumers expect algorithmic game theory and computational social choice of (AI) robots to be attributed to replace some office human workers' tasks. Such as (AI) systems tat address the economic and social computing dimensions of (AI), such as how systems can handle potentially incentives, including self-interested human participants or firms, and the automated (AI) -based agents representing them, e.g. complex or simple office administrative tasks, e.g. typing, accounting, filing, etc. general office tasks which need human office workers who use computers to work to be replaced by (AI) robots to do. So,

Asia office (AI) robots users who expect any office human administration tasks can be replaced by (AI) robots to do.

In Asia computer vision market, Asia computer vision (image analytics), users expect (AI) robots can be replaced to human computer image workers to shorten time to work, or raise image quality to be more clear in the process of pulling relevant information from an image or sets of images to advanced classification and analysis. Such as hospital or clinic x-ray image vision (AI) robot invention, photo image (AI) robot invention etc. any Asia image industry (AI) robot market need.

In Asia collaborative systems work with human (AI) robot market, Asia autonomous systems robots users who expect (AI) robots can be applied its models and collaborative systems to help them to develop autonomous systems that can work collaboratively with other systems and with humans, e.g. car manufacturing, computer manufacturing or any machine manufacturing products. So, Asia machine related manufacturing product industry manufacturers who expect to apply (AI) robots who can assist factory human manufacturing workers to manufacture any products in the short time efficiently.

In Asia language teaching (AI) robot market, language educators expect (AI) robots own natural language processing ability, algorithms that process human language input and convert it into understanding representation, such as Asia translation education market (PWC).

Thus, Asia language teaching businessmen expect (AI) brain invention which needs to be designed to own these above abilities of (AI) robot's manufacturers expectation to provide to them to use from any (AI) robots import.

Is the brain a good model for machine intelligence?

The actions of a human " computer" using paper and pencil to perform a calculation (as the world meant), into a formalized machine, manipulating symbols on an infinite paper tape. But I believe it still has challenges to influence human brain can be invented to machine intelligence successfully. I think (AI) scientists need to solve these further challenges to influence (AI) brain invention success. The limitations include such as below:

Computation is based on functions of integers is limited. They must let computer data can change to words, then words can change images, then any images can change to storage to let (AI) brain to remember to do any analytical and judgement able tasks to make any actions in the short time finally. It is (AI) brain behavioral and analytical mind speed limitation.

Moreover, anther limitation concerns biological systems clearly difference, they must respond to varied stimuli over long period of time, those responses any changes alter their environment and subsequent stimuli. The individual behaviors of social insects, for example, are affected by the structure of the home, they build, and the change their behaviors. Nowadays, (AI) brain invention is called computational neuro science, which have assured that the brain is a computer, it means a machine that is algorithms and architectures. Second, neuro science findings may validate the energy ability of existing algorithms being integral parts of a general (AI) system. It means (AI) robots change adapting environment limitation. It means how (AI) brains adapt to choose to make any analytical mind as well as how to be influenced by external environment to do their behaviors or actions in the efficient way, e.g. how to manufacture many cars efficiently in one factory in the short time.

Consequently, to solve these limitations, we need to know how to apply correct conceptual knowledge to let (AI) robots to learn, e.g. for example, if we know how conceptual knowledge was formed from perceptual inputs, it would crucially allow for the meaning of symbols in an artificial language system to be grounded in sensory " reality". When (AI) scientists can achieve how to solve all above limitation challenges, then (AI) brain invention will achieve more easily.

Must Developed And Developing Countries
Different Artificial Intelligent Development Stages

Must developed and developing countries need artificial intelligent development? If one developed country, e.g. US, UK , Japan , Singapore it does not continue to develop artificial intelligence, robotic, then what disadvantges or weaknesses , it will encounter to compare when it chooses to continue to develop this artificial intelligent technology in society. If one developing country, e.g. China, Korea, Taiwan, it does not continue to develop artificial intelligence, robotic, then what disadantages or weaknesses, it will also encounter to compare when it chooses to continue to develop this artificial intelligent technology in in society. I shall explan the reasons why the results may cause to either the developed country, or the developing country as below:

● How AI help developing countries to communication and agriculture and learning and medical delivery development
Why can AI help developing countries ? Drones that pick inaccessible crops and mobile phones that give medical advice are two of the ways

AI can transform life in the developing world. Artificial intelligence (AI) may improve the lives of the world's poor, the technology needed to revolutionise inefficient, ineffective food and healthcare systems in developing countries is well. For example, in low-income areas, agriculture and healthcare are two critical ecosystems that we can apply AI to immediately; this is not the far future, or even in five years.

Artificial intelligence (AI) has seeped into the daily lives of people in the developed world. From virtual assistants to recommendation engines, AI is in the news, our homes and offices. There is a lot of potential in terms of AI usage, especially in humanitarian areas. The impact could have a multiplier effect in developing countries, where resources are limited.

Emergency Response to developing countries' earthquake natural damage suddence occurrence predicting

AI and machine learning are still finding importance in emerging markets, but certain applications have emerged and are now widely used. For instance, predictive models for disaster relief enable first responders to automatically analyze large-scale behavior and movement through multiple sources of data including social media platforms, web forums, news sources, etc. Based on collected data, responders can scale reconstruction efforts and distribute supplies in a timely manner.

Why and how AI can assist farmers to predict when the earthquake occurs suddenly in order to avoid or reduce the natural damage to their agriculture productive number loss. For example, In 2015, when a major earthquake hit Nepal, more than 8 million people were affected. During the aftermath, drones were used to map and assess the destruction and speed up the rescue mission. The town of Sankhu, situated about 20 kilometers northeast of Kathmandu, was among the highly affected locations. In May 2018, my company Fusemachines and GeoSpatial Systems partnered with Sankhu's city officials to use drones and artificial intelligence in an effort to automatically estimate the reconstruction need. After processing data accumulated from a drone-powered aerial mapping of the region, the team fed this data to advanced machine learning algorithms. Combining drone imagery, digital mapping and machine learning, the team configured region modeling and infrastructure development with higher accuracy. Another organization known as One Concern, a California-based startup, has created a predictive AI program called Seismic Concern to accurately predict seism and is also working on solutions for wildfires, floods and hurricanes.

Smart AI Agriculture

Another application of AI in developing countries is smart agriculture. Farmers monitor crops more effectively and make better predictions on planting, weeding and harvesting using AI tools. It can also be used to analyze one plant at a time and add pesticides only to infected plants and trees instead of spraying pesticides across large swaths of crops. One California-based tech company is an example of this use of AI. So, the developing countries farmers in rural parts of India are also using AI to increase yields through better access to information about the farming season than they would normally have. Technology-enabled process automation offers the agribusiness industry the chance for remarkable growth -- not only in developed countries but around the world. There's a unique opportunity to increase yields, cut down labor costs and improve people's health.

Medicine Delivery to developing countries' patients urgent need

Companies are also leveraging AI to improve access to health care in some of the most remote areas of the world. In Rwanda, for example, Zipline is using drones to deliver medical supplies and blood to hospitals and clinics that are difficult to access by car. This has dramatically impacted people living in remote parts of the country because they are able to get medical help when needed. The drone system in Rwanda has also helped reduce waste of blood by 95%, as noted by Zipline. One Concern has created an AI program called Seismic Concern that accurately predicts seismic events and is also working on solutions for floods, wildfires and hurricanes. The medical field may actually benefit the most from emerging technologies in developing countries.

Assistance to reduce teaching work workload or psychological pressure to teachers in developing countries' schools

Another vital area benefiting from innovative technologies like AI is education. Advanced technologies can enhance how we learn, teach and perform tasks. In most developing countries, schools lack experienced teachers and resources to enhance students' knowledge. As a result, many students still have to walk long distances to get to the nearest school, which has created education gaps, especially in rural areas. AI tools such as personalized learning assistants can simplify learning by making tutoring services and learning materials accessible to all students, wherever they are. Machines can be automated to help students learn basic concepts without a tutor, which companies like Carnegie Learning are working on. This would allow students to learn at any time from anywhere. With AI, education is made easy and accessible to more people.

The initial usage of AI in developing countries has been at a micro level -- solving small, specific problems in a defined industry. As machine learning advances and there is a higher utilization of AI, we will see more complex issues being targeted and resolved. When duly adopted, AI can positively impact future developing countries people everyday lives not just in disaster intervention, education, health care and agriculture but can also help in mitigating poverty, malnutrition and pollution. Especially, in developing nations, to leverage AI's true potential and create a snowball effect. Startups are defining a holistic and humanitarian approach to building more sophisticated, AI-ready societies. Stakeholders in the AI landscape should understand the strengths and nuances of the developing world as well as the limitations of AI and create localized solutions and applications.

Why does smart phone help developing countries communication ? Internet Seen as Positive Influence on Education but Negative on Morality in Emerging and Developing Nations. Internet access differs substantially across the 32 emerging and developing countries polled, with the lowest rates of internet use in South Asian and sub-Saharan African nations. Within countries, computer owners, young people, the well-educated, the wealthy and those with English language ability are much more likely to access the internet than their counterparts. To access the internet, people increasingly use smartphones rather than more cumbersome fixed landline connections and computers. Around the world, both smartphones and basic-feature phones alike are used for sending messages and taking pictures.

In fact, many developing countries young people, students are popular to use smart phones for internet usage aim, instead of communication. Moreover, many developing countries working people are also popular to use smart phones for any working usage in their working time , even non working time any time. So, smart phones (AI) phones will be important communication or leisure tools to developing countries people in the future. Unless, it is one day, scientists can develop another new communication tool to replace smart phones. So, artificial intelligence will be important to influence developing countries people , how to improve or bring positive learning attitudes to students in their daily learning lives. as well as how to raise developing countries people, how to raise working people efficiency or improve performance in their daily working lives. So, AI may bring positive learning or working attitudes to developing countries working people and students both.

The Positive Impact of Mass Media in Developing Countries

Radio, newspapers, television, Internet, social media, etc., all of these are forms of mass media. Each of these outlets has the capability of bringing information to thousands of people with one device. While in some communities it is easy to take advantage of these communication outlets such as television and Internet access, not everyone has access to such outlets. Radio is one of the most common forms of mass media in developing countries because it's affordable and uses less electricity than many other forms of mass media, but only approximately 75 percent of people in developing countries have access to a radio, and roughly 77 percent of people in rural areas have access to electricity.

For developing countries that have implemented forms of mass media in their communities, there have been numerous positive outcomes are influenced to impact developing countries mass media by artificial intelligence as below:

When AI is participated to developing countries mass media, it can influence any radio, television audiences raise more attention to each other through social media platforms such as Facebook and Twitter and create, organize and initiate street protests and campaigns. Furthermore, having access to social media in developing countries, people are able to connect to those that they usually wouldn't have the chance to talk to. Moreover, AI Provides educational opportunities- In many countries, the division between local and national languages as well as issues of literacy can make communication difficult. With the use of mass media, a bridge can be built between these two gaps. In India, there is a radio station that provides information in local languages and respects local culture and traditions. One of the main ways is to create public awareness of what is going on with businesses and government officials. The media plays an important role in giving people the opportunity to act against injustice, oppression and misdeeds that they otherwise wouldn't know about. Information on available healthcare, a mass radio broadcast was sent out encouraging parents to seek treatment at local healthcare facilities for their sick children. With this mass outreach on healthcare, the encouragement of people to take their children to healthcare facilities saved thousands of lives. This easy way of encouraging others and bringing awareness about certain diseases was made possible through a simple radio broadcast. Finally, when AI is participated to media, it may bring many social issues to life that otherwise would remain unknown to many people. In developing countries and communities like Burkina Faso, when the radio broadcast was released about malaria, diarrhea and pneumonia, people were educated and moved to action and knew to take their children to healthcare facilities for preventative care. As it is seen, having access to different media outlets is vital for those in developing countries. Here are three ways that those in developing countries can implement mass media to help their people and communities.

When AI is participated to any internet radio or internet newspaper mass online listening or reading channel. It can provide online radios or newspapers in public places- By providing online radios and newspapers in public areas it gives community members to access news, information and emergency warnings. Even though radios can be on the cheaper side, there are still many people that can't afford to have a radio in their home. By providing one in a local place, not only would it better educate the community members but also it will bring the community together. So, it can make media outlets a two-way platform- Creating a two-way platform between the community and those who are behind the radio stations, newspapers or broadcasts makes the community feel involved and that their voices are being heard. An organization called Soul City in sub-Saharan Africa is showing how well two-way platforms work by engaging their listeners and having them contribute thoughts and ideas about complex issues. Because developing countries radio listening audiences or newspaper readers are popular to accept computer online radio listening channel or online newspaper reading channel to replace traditional paper newspapers or radio machines. So, AI may raise their listening news or reading news leisure feeling from online mass media channel in the future.

● Why do developed countries need to develop AI

Artificial intelligence, or AI, is driving massive shifts across the globe, and every day more questions arise. What impact will AI have on the workforce and how can we prepare for it? How can we encourage economy-boosting and job-creating technologies? How can we ensure that AI will be implemented ethically and with minimal bias? How will society benefit? For developed country, such as US example. None of the US, Israel and Russia have a formal national AI policy yet. Private sector companies such as Google, Amazon and Apple and the US department of defence are driving the bulk of AI investment in the United States. Though Israel does not have a specific policy, it is keenly focused on AI and has seen the number of AI start-ups triple since 2014.

Developed country may learn whether what weakness it is lacking when it does not continue to develop AI from one another developed country. Which countries are approaching AI most effectively, and to what degree is there opportunity for greater international collaboration? It may be too early to tell; however, when analyzing the best practices of existing national AI policies, there is much that can be learned. These are the specific areas to consider. When one developed country continue to develop or research AI, it may bring these benefits as below:

On gathering Data aspect, from self-driving vehicles to smart cities, data is the driver behind AI. Innovation in the United States is limited without a national strategy that answers questions about protocol and ownership. France and Denmark, on the other hand, are opening government data. France is hosting troves of centrally collected public and private data that it plans to make available as part of its strategy. Conversely, by taking a restrictive position on issues of data collection (as indicated by the implementation of General Data Protection Regulation), the EU is putting manufacturers and software designers at a disadvantage while balancing the demand for privacy. On raising technologica talent aspect, the demand for AI talent far outweighs the available supply. As a result, almost every nation's strategy addresses talent development. Canada's AI strategy is distinct in that it primarily focuses on research and talent strategy. The country boasts AI degree programmes and is building a $127 million research facility in Toronto. Companies like Facebook and my own company, Uptake, are investing in Canada to access this talent pool. On AI legal technological innovation aspect, a whole host of legal questions swirl around AI. The country is developing a bill for AI liability that will be ready in March 2019. The government hopes the legal framework will attract investors by providing a simple, comprehensive guideline to enable the broad use of AI systems. So, when the developed country applied AI technology to assist any lawyers to work, then AI can help them to reduce the workload to draft any legal documents more easier. So, any developed countries lawyers' draft legal documents time must reduce if the developed countries lawyers accept to apply AI to assist their legal works. One of the great promises of AI is its potential for improving quality of life. But without the right planning and oversight, we risk exacerbating problems of inequality or marginalizing groups of people. As an example, India's AI strategy is focused on leveraging the technology not only for economic growth, but also for social inclusion.

AI may bring what benefits to developed countries

From SIRI to self-driving cars, artificial intelligence (AI) is progressing rapidly. While science fiction often portrays AI as robots with human-like characteristics, AI can encompass anything from Google's search algorithms to IBM's Watson to autonomous weapons. Artificial intelligence today is properly known as narrow AI (or weak AI), in that it is designed to perform a narrow task (e.g. only facial recognition or only internet searches or only driving a car). However, the long-term goal of many researchers is to create general AI (AGI or strong AI). While narrow AI may outperform humans at whatever its specific task is, like playing chess or solving equations, AGI would outperform humans at nearly every cognitive task.

Why research AI safety? Would AI bring war when AI is continued to develop by developed countries? In the near term, the goal of keeping AI's impact on society beneficial motivates research in many areas, from economics and law to technical topics such as verification, validity, security and control. Whereas it may be little more than a minor nuisance if your laptop crashes or gets hacked, it becomes all the more important that an AI system does what you want it to do if it controls your car, your airplane, your pacemaker, your automated trading system or your power grid. Another short-term challenge is preventing a devastating arms race in lethal autonomous weapons.

In the long term, an important question is what will happen if the quest for strong AI succeeds and an AI system becomes better than humans at all cognitive tasks. As pointed out by I.J. Good in 1965, designing smarter AI systems is itself a cognitive task. Such a system could potentially undergo recursive self-improvement, triggering an intelligence explosion leaving human intellect far behind. By inventing revolutionary new technologies, such a superintelligence might help us eradicate war, disease, and poverty, and so the creation of strong AI might be the biggest event in human history. Some experts have expressed concern, though, that it might also be the last, unless we learn to align the goals of the AI with ours before it becomes superintelligent.

There are some who question whether strong AI will ever be achieved, and others who insist that the creation of superintelligent AI is guaranteed to be beneficial. At FLI we recognize both of these possibilities, but also recognize the potential for an artificial intelligence system to intentionally or unintentionally cause great harm. We believe research today will help us better prepare for and prevent such potentially negative consequences in the future, thus enjoying the benefits of AI while avoiding pitfalls.

How can AI be dangerous when developed countries continue to develop AI to become weapon to replace soldiers?

Most researchers agree that a superintelligent AI is unlikely to exhibit human emotions like love or hate, and that there is no reason to expect AI to become intentionally benevolent or malevolent. Instead, when considering how AI might become a risk, experts think two scenarios most likely:

The AI is programmed to do something devastating: Autonomous weapons are artificial intelligence systems that are programmed to kill. In the hands of the wrong person, these weapons could easily cause mass casualties. Moreover, an AI arms race could inadvertently lead to an AI war that also results in mass casualties. To avoid being thwarted by the enemy, these weapons would be designed to be extremely difficult to simply "turn off," so humans could plausibly lose control of such a situation. This risk is one that's present even with narrow AI, but grows as levels of AI intelligence and autonomy increase.

The AI is programmed to do something beneficial, but it develops a destructive method for achieving its goal: This can happen whenever we fail to fully align the AI's goals with ours, which is strikingly difficult. If you ask an obedient intelligent car to take you to the airport as fast as possible, it might get you there chased by helicopters and covered in vomit, doing not what you wanted but literally what you asked for. If a superintelligent system is tasked with a ambitious geoengineering project, it might wreak havoc with our ecosystem as a side effect, and view human attempts to stop it as a threat to be met. So, a super-intelligent AI will be extremely good at accomplishing its goals, and if those goals aren't aligned with ours, we have a problem. You're probably not an evil ant-hater who steps on ants out of malice, but if you're in charge of a hydroelectric green energy project and there's an anthill in the region to be flooded, too bad for the ants. A key goal of AI safety research is to never place humanity in the position of those ants.

Why the recent interest in AI safety ?

Stephen Hawking, Elon Musk, Steve Wozniak, Bill Gates, and many other big names in science and technology have recently expressed concern in the media and via open letters about the risks posed by AI, joined by many leading AI researchers. The idea that the quest for strong AI would ultimately succeed was long thought of as science fiction, centuries or more away. However, thanks to recent breakthroughs, many AI milestones, which experts viewed as decades away merely five years ago, have now been reached, making many experts take seriously the possibility of superintelligence in our lifetime. While some experts still guess that human-level AI is centuries away, most AI researches at the 2015 Puerto Rico Conference guessed that it would happen before 2060. Since it may take decades to complete the required safety research, it is prudent to start it now.

Because AI has the potential to become more intelligent than any human, we have no surprise way of predicting how it will behave. We can't use past technological developments as much of a basis because we've never created anything that has the ability to, wittingly or unwittingly, outsmart us. The best example of what we could face may be our own evolution. People now control the planet, not because we're the strongest, fastest or biggest, but because we're the smartest. If we're no longer the smartest, are we assured to remain in control?

A captivating conversation is taking place about the future of artificial intelligence and what it will/should mean for humanity. There are fascinating controversies where the world's leading experts disagree, such as: AI's future impact on the job market; if/when human-level AI will be developed; whether this will lead to an intelligence explosion; and whether this is something we should welcome or fear. But there are also many examples of of boring pseudo-controversies caused by people misunderstanding and talking past each other. When one developed country continue to develop AI, can itself country's all factories workers will lose jobs, due to AI can replace them to do simple works in factories, or any public transport drivers, e.g. bus drivers, ferry , tram, train drivers, they will lose jobs, when AI (non manual driving drivers) can replace all public transport drivers. So, some occupations will lose if developed countries continue to develop or research AI to replace human to do some simple jobs, such as some cooking jobs can be done by AI. So, it is possible that future cookers won't be needed, because AI cooking skills may be better than them to cook any good taste chinese or western food in restaurants. If you drive down the road, you have a subjective experience of colors, sounds, etc. But does a self-driving car have a subjective experience? Does it feel like anything at all to be a self-driving car? Although this mystery of consciousness is interesting in its own right, it's irrelevant to AI risk. If you get struck by a driverless car, it makes no difference to you whether it subjectively feels conscious. In the same way, what will affect us humans is what superintelligent AI does, not how it subjectively feels.

In fact, AI may be make any brokers jobs in financial market. the main concern of the beneficial-AI movement isn't with robots but with intelligence itself: specifically, intelligence whose goals are misaligned with ours. To cause us trouble, such misaligned superhuman intelligence needs no robotic body, merely an internet connection – this may enable outsmarting financial markets, out-inventing human researchers, out-manipulating human leaders, and developing weapons we cannot even understand. Even if building robots were physically impossible, a super-intelligent and super-wealthy AI could easily pay or manipulate many humans to unwittingly do its bidding. So, future brokers will be replaced by AI, when AI can be made to own financial brokers' analytical mind to make more accurate whether the share price will rise up or fall down to compare human financial brokers' analytical mind. The robot misconception is related to the myth that machines can't control humans. Intelligence enables control: humans control tigers not because we are stronger, but because we are smarter. This means that if we cede our position as smartest on our planet, it's possible that we might also cede control.

Not wasting time on the above-mentioned misconceptions lets us focus on true and interesting controversies where even the experts disagree. What sort of future do you want? Should we develop lethal autonomous weapons? What would you like to happen with job automation? What career advice would you give today's kids? Do you prefer new jobs replacing the old ones, or a jobless society where everyone enjoys a life of leisure and machine-produced wealth? Further down the road, would you like us to create superintelligent life and spread it through our cosmos? Will we control intelligent machines or will they control us? Will intelligent machines replace us, coexist with us, or merge with us? What will it mean to be human in the age of artificial intelligence?

Why do developed countries people need AI ?

Why do we assume that AI will require more and more physical space and more power when human intelligence continuously manages to miniaturize and reduce power consumption of its devices. How low the power needs and how small will the machines be by the time quantum computing becomes reality? Why do we assume that AI will exist as independent machines? If so, and the AI is able to improve its Intelligence by reprogramming itself, will machines driven by slower processors feel threatened, not by mere stupid humans, but by machines with faster processors? What would drive machines to reproduce themselves when there is no biological incentive, pressure or need to do so?

Who says superior AI will need or want to have a physical existence when an immaterial AI could evolve and preserve itself better from external dangers. What will happen if AI developed by competing ideologies, liberalism vs communism, reach maturity at the same time, will they fight for hegemony by trying to destroy each other physically and/or virtually. If AI is programmed to believe in God, and competing AI emerges programmed by muslims, christians or jews, how are the different AI's going to make sense of the different religious beliefs, are we going to have AI religious wars? What if the "powers that be" greatest fear is the emergence of a super AI that police's and rationalizes the distribution of wealth and food. A friendly super AI that is programmed to help humanity by, enforcing the declaration of Human Rights (the US is the only industrialized country that to this day has not signed this declaration) ending corruption and racism and protecting the environment.Most benefits of civilization stem from intelligence, so how can we enhance these benefits with artificial intelligence without being replaced on the job market and perhaps altogether?

Key to the process of machine learning are neural networks. These are brain-inspired networks of interconnected layers of algorithms, called neurons, that feed data into each other, and which can be trained to carry out specific tasks by modifying the importance attributed to input data as it passes between the layers. During training of these neural networks, the weights attached to different inputs will continue to be varied until the output from the neural network is very close to what is desired, at which point the network will have 'learned' how to carry out a particular task. A subset of machine learning is deep learning, where neural networks are expanded into sprawling networks with a huge number of layers that are trained using massive amounts of data. It is these deep neural networks that have fuelled the current leap forward in the ability of computers to carry out task like speech recognition and computer vision.

In conclusion, when developed countries continue to develop AI, it may bring positive advantages to bring raising productivies, or efficiencies, but it may also raise unemployment ratio to any low skill or low knowledge jobs in ther societies. However, human future society will need to change to be better to raise our living standard. But AI is one kind the best choice tool to achieve this aim in our future, so I agree developed countries continue to develop or research AI to be the super -human machine.

Reference

PWC, Sizing the prize, see: https:// www. pwc. com/gx/en/issues/data- and- analytics/publications/artificial- intelligence- study.html.

Rotman M. J., Data mining-a practical approach to database marketing (1995). IBM.

CHAPTER IV

How robots help manufacturers to raise productivity

Some scientists explain that artificial intelligence means which is an expert system, computer software that embodies a portion of the specialized knowledge of a human portion in a specific, narrow domain, owns decision making ability of human expert. The (AI)technology is based on the premise that what makes a person an expert is years of experience that enables who recognizes certain patterns in a problem as being similar to pattern. For example, in the future artificial intelligence system can be applied to control air traffic, design to computer configuration, medical diagnosis, instruction/training, speech/interpretation, monitoring to (nuclear plant), planning to mission, factory scheduling, prediction weather, repairing telephone, automatic driving etc. different industries.

Artificial intelligence characteristics include: creative, adaptive , common sense, fact processing, quick replication, broad focus permanent and consistent skill. Otherwise, traditional computer expert system disadvantage includes perishable, unpredictable, slow reproduction, expensive, slow reproduction, slow processing lacks inspiration, needs instruction, narrow focus only machine knowledge. So, artificial intelligence is a branch of computer science devoted to creating computer to influence software and hardware to attempt to create human intelligence or human intelligent behavior. It is learning from experience, responds flexibility in situation that are, new or not anticipated.

Thus, (AI) can be learnt programmed knowledge to solve problems, using reasoning in solving problem, understanding and inferring facts and rules, recognizing the relative importance of different elements in a situation. In summary, artificial intelligence is concerned with two basic ideas mainly: The first idea, it involves studying the thought processes of humans to understand what intelligence is; the second idea, it deals with representing thought processes using companies to create artificially intelligent entities for testing the theories of intelligence.

It brings this question:

Can AI apply these strengths to assist human to raise productivity?

● Can (AI) impact human job nature?

Human need concern these two questions:

Will artificial intelligence (AI) reduce some human jobs in order to instead of replacing machines to do?

If (AI) can replace human some jobs, does it reduce productivity or raise productivity ?

Due to artificial intelligence is the ability of machines to do thing, that people would require intelligence. For example, artificial intelligence machine man driving(self-driver), it (AI) machine man driving research is an attempt to discover and describe aspects of human intelligence that can be simulated by driving machine functions. Alternatively, (AI) mathematical research may be another viewed as an attempt to develop a mathematical theory function to describe the abilities and actions of things (natural or man-made) exhibiting intelligent behavior and server as a design of intelligent calculation machine function.

Can artificial intelligence machines better than manual to replace traditional human service job? For example, can artificial intelligence machine man (self-driving) driver drive to replace human driver?

Is (AI) replaced to human any job, Can it bring positive impact to raise more productivity than human ?

I shall compare the differences between humans and computers function as below:

The characteristics of humans are good at recognizing various things, either seen before or not, recognizing the relationship patterns between things. Human thinking is common sense reasoning, combining all types of sensory input, acting appropriately in novel situations, learning new things and changing behavior patterns, making decisions , even when given incomplete information, working with noisy, incomplete information gathering behaviors . However, characteristics of computers are good at: The tasks humans do naturally are extremely difficult for a computer program as intelligent, which must be able to do the same kind of tack as humans do naturally.

Hence, it seems that (AI) can raise productivity, due to it is manufactured computer software and computer software can do complex jobs to compare human, even human can not do these complex jobs.

Hence, (AI) is an combination of many different success and technologies: Linguistics - computational and socio, philosophy-logic, philosophy of mind and of language, electronic engineering -image and speech processing, pattern recognition, robotics, machine learning, neural networks, optimization scheduling, management information system and decision making. So, it is possible that (AI) can impact human job nature to raise more productivity to do better than human working behavior in the future.

How can (AI) influence labor market?

● How can human society job nature

to be changed to artificial intelligent society?

From the first intelligent perspective reason view point, artificial intelligence is making machines " intelligent" acting as humans expect people to act. Artificial intelligence has ability to distinguish computer responses from human responses, it owns knowledge to solve expert problem. From another research perspective reason view point, artificial intelligence is the study of how to make computers do things which, at the moment, people do better (Rich & Knight, 1991, p.3).

(AI) researchers are native in a variety of domains, e.g. formal tasks (mathematics, games), tasks (perception, robotics, natural language, common sense reasoning), expert tasks (financial analysis, medical diagnostics, engineering, scientific analysis and other areas).

From the second business perspective reason view point, (AI) is a set of many powerful tools, and methodologies for using those tools to solve business problems. From a programming perspective reason view point, (AI) includes the study of symbolic programming problem solving and search .

From the third human technological perspective reason view point, today's computer can do many well-defined tasks, for example, arithmetic operations, are much faster and more accurate than human beings. However, the computers' interaction with their environment is not very sophisticated yet. How can human test whether a computer has reached the general intelligence level of a human being? Can a computer convince a human interrogator that it is a human? But before thinking of such advanced kinds of machines, human will start developing our own extremely simple " intelligent" machines.

So, it is possible that human society job efficiency and performance and productivity will to be changed better, due to artificial intelligent robots are applied to assist human to do anything when (AI) technology is developed to the mature stage in the future.

● Why does human need artificial intelligence machines?

One of major division in (AI) is between humans who think (AI) is the only serious way of finding out how we (human) work and human who want companies to do very smart things, independently of how we (human) work. This is the important distinction between cognitive scientists vs engineers.

One of another major division in (AI) is between symbolic (AI), which represents information through symbols and their relationships. Specific Algorithms are used to process these symbols to solve problems or deduce new knowledge and connectionist. So (AI) , which represents information in network. Biological processes underlying learning, task performance and problem solving are imitated from human mind behaviors.

Thus, it is possible that artificial intelligence machines can do the better human task behavior to achieve productivity and efficiency raising.

● How does artificial intelligence influence future working changing in automation employment and productivity aspects?

(AI) influences the automation working changing aspect, as companies increasingly use robots on production lines or algorithms to optimize their logistics manage inventory, any carry out other core business functions. Technological advances are creating a new automation age in which ever-smarter and more flexible machines will be deployed on an ever larger scale in the marketplace. However, researching artificial intelligence with how influences human working nature. We need to answer these questions:

How will automation transform the workplace?

What will the implications for employment?

What is likely to be its impact both on productivity in the global economy and on employment?

Advances in robotics, artificial intelligence, and machine learning are growing in a new age of automation as machines match or outperform human performance in a range of work activities, including ones requiring cognitive capabilities.

What factors are determined the changing in workplace adoption by artificial intelligence innovation?

What advantages are automation?

Automation of activities can be enabled businesses to improve performance by reducing errors and improving quality and speed, and achieving outcomes that go beyond human capabilities.

Some scientists indicated based on their scenario modeling. They estimated automation could raise producing growth globally by 0.8 to 1.4 percent annually. Almost, the activities people are paid almost $16 trillion in wages to do in global economy have the potential to be automated by adopting currently demonstrated technology. According to their analysis of more than 2,000 work activities across 800 occupations. When less than 5% of all occupations have of least 30% of activities that could be automated. They also indicated that technical economic and social factors will determine automation. Continued technical progress, for example, in areas such as natural language processing is a key factor beyond technical feasibility , the cost of technology, competition with labor including skills, and supply and demand dynamics, performance benefits including and beyond labor cost savings and social and regulatory acceptance will affect (alter) the scope of automation.

Other some scientists also indicate U.S. country for example, the anticipate shift in the activities in labor force of a similar order of magnitude as the long term sight away from agriculture and decreases in manufacturing. Share of employment in the United States both which were achieved. So, those factors can influence why artificial intelligence technology needs. So, it is possible that future agriculture and manufacturing both industries will apply (AI) technology manufacturer-kind of job nature to raise productivity instead of farmers, fruit picking workers, farming transportation labours as well as factory manufacturing workers and supervisors etc. human-kind of job nature.

● Is artificial intelligence possible to replace labor ?

Not just intelligence, but also debating, if machines are capable of having a conscious minds. Artificial intelligence has those characteristics as below:

On functionalism aspect, artificial intelligence inputs mental states, sensory inputs, (beliefs, desires being in pain feeling) and behavioral outputs. Since mental states are identified by a functional role, which are thoughts to be manifested in various systems. Even, perhaps computers which are physical devices with electronic substrate that inform computations on inputs to give outputs similar to brains which are artificial intelligence composed of part any intrinsic relationship to each other. Thus, artificial intelligence activities is not the whole itself, but into parts or on external influence on the parts.

On dualism aspect, artificial intelligence is a set of views about the relationship between mind are matter. On materialism aspect, it builds the only thing that exists is matter, including consciousness.

On biological naturalism aspect, it is similar a human brain than feels pains makes mental situation. So, artificial intelligence is similar biologist which might to be excited to human labor work. Hence, it seems artificial intelligence can change (alter) or replace human labor work of nature in order to raise efficiency and productivity in possible any working environment when it own human mind and effort absolutely.

● Can (AI) technology replace human labour nature of work to raise better productivity and efficiency?

On technological innovation reason view point, the history development of artificial intelligence studying the intelligence is one of most ancient scientific discipline. The history development of artificial intelligence what aims to achieve human use to sense, learn remember and think, logic probability, decision making and calculation develop from mathematics, instead of replacement human labor functions.

Artificial intelligence history development aim is the scientific analysis of skills in connection and practice with the appearance of computers from 1950 year beginning. The artificial intelligence (AI) can deal with the ultimate challenges. How can (either biological or electronic) mind sense, understand and manipulate a world that is much simple and more complex than itself? And what if would human like to construct something with such capabilities?

The general-purpose software of the early period of (AI) were only able to solve simple tasks effectively and failed when which should be used in a wider range or an more difficult tasks. One of the sources of difficulty was that early software had very few or mix knowledge about the problems which handled, and activities successes by simply syntactic manipulation. Moreover, the other difficulty was that many problems that were tried to solve by the (AI) were untreatable.

The early (AI) software whether trying step sequences based on the basic facts about the problem that should be solved, experimented with different combinations till which found a solution. From the end the 1960 year, developing the so-called expert systems were emphasized. These systems had (sue-based) knowledge base about the field which handled. Till to the beginning of the 1970 year, (Prolog) the logical programming language was born, which was built in the computation realization of a version of the resolution calculus. (Prolog) is a remarkably prevalent tool in developing expert systems (on medical, judiciary and other scopes), but natural language parsers were implemented in this language. Then, in 1981 s, the Japanese announced the fifth generation computer system project, a 10 years plan to build an intelligent computer system that use the (Prolog) language as a machine code. Nowadays, (AI) can be applied any industries, such as car manufacturing industry can use (AI) technological machine-men manufacture car, instead of replacing human labors in factory. Even, in the future, using (AI) machine-men drivers can drive any private cars or public transportation tools, instead of replacing human drivers, e.g. bus, train, tram, ferry etc. Also in the future, machine-men can replace housewives to serve families to do housekeeping clean job , e.g. cleaning toilets, bathrooms, kitchens, even cooking functions at home. So (AI) machine-man can reduce housewives works at home. Moreover, (AI) machine man can take care old people , when who are living at homes or elder care centers.

So, it seems artificial intelligence (AI) will be possible developed to manufacture a new generation machine-man to assist (serve) families to do any simply cleaning or cooking jobs at homes. Moreover, the overall demand of (AI) general social needs will also rise, such as security, driving transportation tools, restaurant cleaning, elder centers care service etc. (AI) will have chance often to practise to do its tasks every day. It is possible that it can improve worker individual simple job duty to raise productivity and efficiency to be fast.

● Why can artificial intelligence satisfy human needs?

First, On machine-man satisfactory demand aspect view point, it makes computers that think, it is the automation of activities. We associate with human thinking: like decision making, learning. It is the act of creating machine that perform function that require intelligence when performed by people. It is the study of mental faculties through the use of computational models. It is the study of computations that make it possible to perceive, reason and act. It is a branch of computer science that is concerned with the automation of intelligent behavior. It is anything in computing service that human don't yet know how to do property.

Second, on thought aspect artificial intelligence means systems thank think like humans, systems that think rationally.

Third, on behavioral aspect, artificial intelligence systems that act like human and that systems act rationally. However, the basic objective of (AI) is to represent human's thought processes in computation . These machines are supposed to exhibit behavior that. It is performed by a human being, would be considered intelligent. However, some authors feel (AI) has disadvantages, such as it is not creative, it is excited in the use of sensory devices, it can't make use of a very wide context of experiences and it does not use common sense.

For speech recognition and understanding function needs example, (AI) can be applied in speech recognition and understanding function, which (AI) speech or voice recognition is a data input method. For example, the computer recognizes and understands one (or a few) word commands. Speech understanding on the other hand is the computer's ability to understanding a spoken language. That is , the computer understands the meaning of sentences, an paragraphs through (AI).

So, (AI) can be attempted to learn human language how to speak. It is similar to translate human language skill, instead of actual human speaking skill. Also, (AI) can assist handicap learning or language student how to listen different languages by machine-man sounds from computers more accurately.

So, it seems that it (AI) can replace human language teachers speaking function and can change teaching language nature of job in language speaking and listening education industry. IT will improve its language speaking performance to teach students to raise teaching efficiency and teaching performance to satisfy student learning needs in short time.

● Is artificial intelligence one good choice for human future technological benefit?

Nowadays, new technology development is popular. However, artificial intelligence is one kind of new technology choice among different technologies innovation. So it brings this question: Is artificial intelligence technology value to invest? To answer this question. I shall indicate some other new technology developments to compare (AI) technology development to judge which has urgent needs to achieve human expectation nowadays.

For example, why is green peace interested in new technologies? New technologies features prominently in our ongoing campaigns against genetic modified crops and number power. However, which are also an integral part of our solutions to environmental challenges, including renewable energy technologies, such as solar, wind and wave (water) power energy as well as waste treatment technologies, such as mechanical, biological treatment.

It seems humans need concern how to apply (AI) technology to solve environment pollution challenges in our future. So, environment protective, agriculture, natural energy technology will be popular demand to attempt to apply (AI) technology to solve their challenges or apply (AI) to assist to develop their industry. It can raise farming food and fruit number more than farmers. It seems that it can raise farming productivity.

What is relationship between
(AI) productivity and economy growth?
● How can artificial intelligence technology influence economy?
Advances in artificial intelligence (AI) technology and related fields have opened up new markets and new opportunities progress in critical areas, such as health, education, energy, economic development, social welfare and the environment pollution.

(AI) automation will continue to create wealth and expand the global economy development in the future. However, when many will benefits that growth won't be costless and will be accompanied by changes in the skills, that workers need to increase productivity in the economy and structural changes in the economy. So, in the skills that workers need to succeed in the economy and structural changes.

I shall indicate why aggressive policy action will be needed to help Americans who are disadvantaged by these changes , due to (AI) technology is caused. For automation industry change example, artificial intelligence (AI) capabilities will enable automation of some tasks that have long required human labor. These artificial intelligence technology introduction can increase new opportunities for individuals. The economy and society, but (AI) has also the potential to disrupt be current livelihoods of many Americans. However, (AI) leads to unemployment and increase in inequality over the long run depends not only on the (AI) technology itself, but also on the institutions and policies that are changed.

Thus, it is possible that (AI) technology will raise some countries unemployment number if the employer apply (AI) technology workers to replace human labor in their factories, but it can also raise productivities for these employers, when (AI) can spend less time to compare labor to do any human's same tasks in order to raise productivity.

● Can (AI) influence global economy growth?
Technological progress is main driver of growth of GDP per capita, allowing output to increase faster than labor and capital . However, technology can increase productivity, but also decrease the number of labor hours needed to create a unit of output. So (AI) causes unequal to labor wage decreases, even reduces the number of labor to manufacture, e.g. artificial intelligence technology of automation car manufacturing industry; clothing

manufacturing industry; plane manufacturing etc. high technology of artificial intelligence manufacturing method. But (AI) should be potential environment benefit, although it raises unemployment ratio. Moreover, it can rise production , due to many skilled craft were replaced by the combination of machines and lower-skilled labor. The result of (AI) technology introduction , it causes output per hour risen when inequality declined, driving up average living standards, but the labor of some high-skill workers was no longer as valuable in the market. Otherwise, if (AI) technology is continue developed to be success. Some routine intensive occupations will be loss, which focused on predictable, e.g. easily programmable tasks, such as switchboard operators, filing clerks, travel agents, and assembly line workers would be particularly replaced by new (AI) technology. However, at the same time, (AI) technology development will bring these benefits: improvement in education (training (AI) technology scientists) , due to (AI) manufacturing technology needs are raising to businesses and institutional changes, such as the reduction in unionization and raising in the minimum wage to the (AI) manufacturing technology skilled labor in factories.

Because (AI) technology is not a single technology, but rather a collection of technologies that are applied to specific tasks, the effects of (AI) will be felt unevenly though the economy. It will bring some tasks will be most easily automated than others , and some jobs will be affected more than others, both negatively and positively. Finally, new jobs are likely to be directly created in areas , such as the development and supervision of (AI) as well as indirectly created in a range areas though out the economy as higher incomes lead to expanded demand.

However, if (AI) technology could dominate global labor markets. If labor productivity increases, do not influence to wage increases, then the large economic gains brought about by (AI) technology could be increased wealth inequality, due to employers can reduce production cost, but workers (labors) wages will not be increased, even will be decreased. Hence, it seems the (AI) technology will bring disadvantages to labor market to cause unemployment or reduce wages in possible, although it can reduce employer individual salary (wage) expenditure and it can raise productivity.

● How can artificial intelligence impact global economy growth?
Artificial intelligence (AI) technology is a branch of computer science that aims to create intelligent machines that work and react like humans. So,

(AI) is a technology that appears to impact (influence) human preference by learning, understanding complex contents, enhancing humans in executing both routine and non-routine tasks. In the future, (AI) technology that can be virtual personal assistant, as well as it may exist, such as robots with human-like processing capabilities.

How can (AI) technology impact global economy growth over the next 10 years? During this time period, (AI) technology is predicted to have wide-ranging applications including: Machine learning that automates analytical model building by using algorithms that allow machines to operate without human assistance.

In global education aspect, potential applications include predicting cause-and-effect relationships from biological data, identifying new drugs, self-driving cars, and protecting against fraud, improved natural language processing that allows computers to continue to better analysis, understand and generate language to interface with humans using natural human languages. For example, transcribing notes dictated by physicians, automatically drafting articles and translating text and speech. So (AI) technology can be applied to education aspect to improve humans' knowledge level.

In visual art aspect, (AI) machine vision that allows computers to identify objects, scenes and activities in images. Current applications of (AI) machine vision include providing objective descriptions for the blind seeing(visual) needs.

We except the economic effects of (AI) technology to include both direct GDP growth from sectors that develop or manufacture. (AI) technology and indirect GDP growth through increased productivity in existing sectors that employ some form of (AI). If (AI) technology is an increasingly critical component of more products, it will become an integral part of many people's lives. Thus, (AI)'s ability to influence economic activity, rather than the economic or development status of the region. (AI) has the potential to impact income classes and to bring significant gains to both developed and developing countries. For example, (AI) has the potential to optimize good production around the world by analyzing agricultural regions and identifying what is necessary to improve crop yields.

In estimating the future economic effects by (AI) technology innovation, it is important to note that it is challenging to accurately predict which applications of (AI) will ultimately be commercially successful. In micro level economic influence, we need to apply methodologies to estimate the economic effects of investment in firms developing (AI) technology since investment levels in a technology are a telling sign of the future potential of that (AI) technology.

● How can (AI) influence GDP of high income countries in the next ten years?

How (AI)'s development may affect the global economy over the next ten years. In fact, (AI) technology has the potential to affect business across the global in a wide range of industries in ways only a number of technologies have done in the parts. For example, (AI) technology's expected to be a useful tool for enhancing human capabilities and in some instances replacing functions, such as driving a car, adoption of broadband internet, mobile telephone, industrial robotic automation have served to enhance human capabilities.

However, significant public debate has focused on projections of (AI) technology's effect on the labor force. However, large companies prefer to invest in (AI) technological industry. For example, face book's (AI) research lab., google machine intelligence lab. and micro soft machine learning and artificial intelligence research division are all making advances in (AI) technology and investing in the industry's top talent. Additionally, between 2010 year and 2015 year, nearly $5 billion in venture capital funding invested in firms across the global developing and employing (AI) technology (Facebook (AI) Research).

● How can artificial intelligence impact on workplace?

Modern information technologies and the labor economy growth of machines is powered by artificial intelligence have already strongly influenced the world of work in the 21 ST century. Computers, algorithms and software simplify every tasks and it is impossible to image how most of our life could be managed without them. How can be the information economy characterized by exponential growth replaces the most production industry based on economy of scales? What will the future world of work look like and how long will it take to get? Will the future world of work be a world where humans spend less time earning their livelihood? Alternatively, are mass unemployment, mass poverty and social distortions also possible scenario for the future, where robots, artificial intelligence systems play an increasingly central role? These questions concern how artificial intelligence further development . Can influence labor economy growth on workplace ? When the labor market has widespread impact on intelligence property, information technology, product liability, competition and labor and employment laws.

How (AI) technology impacts on labor workplace.

The future influence any organizations how labor economies use of (AI) can be analyzed, such as deep machine learning is based on a set of model high level data. Unlike human workers, the machines are connected the whole time in workplace. If one machine makes a mistake, all autonomous systems will keep this in mind and will avoid the same mistake the next time.

Over the long run intelligent machines will win against every human expert. Production robots have been replacing employees because of the (AI) technology. They work more precisely than humans and cost loss. Creative solutions like 3D printers and the self learning ability of these production robots will replace human workers, the automatic data recording and data processing, traditional back office activities are no longer in demand. Autonomous software will collect necessary information and will send it to the employee who needs it. Additionally, dematerialization leads to the phenomenon that traditional physical products are becoming software. For example, CD or DVDs are being replaced by streaming services. The replacement of traditional event ticket, e-travel ticket service products or hard cash will be the next step, due to the possibility of payment by smartphone. So, (AI) technology will impact human's daily life consumption behaviors in the future. For another example, transportation tools, such as boats and ferries and private vehicles will use sensors and navigating without human input. Taxi and truck drivers will become obsolete, the stock store applies to stock managers and postal carriers of the delivery is distributed by (AI) machine delivery methods.

What is the relationship between (AI) and (CRM)?

● Can (AI) technology impact on customer relationship management (CRM) to improve service performance ?

Nowadays , (AI) is a technology almost as old as the computer industry itself, it is similar with the advent of personal assistants function to businesses and personal promotion channel, such as (Amazon's Alexa, Apple's Siri, Google's Assistant) image recognition (face book), personalized recommendations (Netflix , Amazon). Those innovations have been driven by a increase in processing power, lower cost hardware, and the exploding creation and availability of data. It seems, (AI) technology can impact global customer service management method.

How to forecast economic impact modeling to (AI) will affect global economy? Can human forecast business revenue growth and job creation (or destruction) based on (AI) applied to customer relationship management (CRM) activities? In addition to the economic impact on (AI) or (CRM) which can include an estimate of the economic impact attributable to sales forces customer base. What can economic benefits be brought to (CRM) from (AI) technology?

Artificial intelligence(AI) comprises a set of technologies that use natural language processing, machine learning, knowledge graphs, and other tools to answer questions, discover insights and provide recommendations. Computer systems can use (AI) hypothesize and formulate possible answers based on available evidence can be trained through the ingestion of vast amounts of content, and automatically adapt and learn from (AI) self mistakes and failures.

So, any business organizations (customer service departments) can provide efficient and effective customer relationship management of excellent customer service quality if which applied (AI) technology system. The different type of (AI) systems include: (AI) system platforms, machine learning (AI) based data preparation and enrichment tools, machine vision/ image recognition, voice speech recognition, text analysis and natural language processing, bots , e.g. face book website and virtual digital assistance solutions, social media pattern analysis , sentiment analysis, advanced numerical analysis (e.g. IOT streaming , machine logs), supporting technologies, knowledge base dialog management, Q&A processing etc. different (AI) technology system customer relationship management (CRM) tools.

(AI) (CRM) of activity can include these categories, such as: corporate marketing, marketing operation, field marketing, customer support, digital commerce, customer analytics, customer influenced product or service design, product or service pricing, finance information, presentation, customer billing, inventory , logistics and fulfilment support, partner management etc. different CRM tools.

(AI) technology of CRM has been carrying on plan different stages to achieve CRM personal assistant tool for businesses. The stages are such as, in the beginning stage of (AI) projects in place, implement now, pilot phase next year in the final stage of (AI) customer relationship management tools are foreseeable future. So, this CRM technology has been improved to plan in different stages every year to prepare to achieve full capacity of CRM service quality for businesses to use in the future.

Hence, how to develop an estimate prediction of the economic impact (AI) technologies could have CRM activities, which depends on gathering macroeconomic information on business revenue and the basic marketing of business revenue and the basic markup of business expenses by major functions (customer support, marketing and sales , production etc.)

An economic impact model that can gather data together and forecast the results how (AI) artificial intelligence technology brings (CRM) customer relationship management benefits to businesses, e.g. surveys investigation includes IT spending by sample countries, GDP and population estimates and forecasts, revenue per employee and ratios of IT spend to GDP. Surveys (questionnaire questions) of forecast results are influenced by (AI) impact can include: results are projected from surveys and rely on estimates are made by respondents on the expected financial improvements in categories of (AI) –assisted customer relationship management activities. The forecast assumes that these estimates are correct; financial estimates are based on estimates of "first year" improvement from full (AI) implementation; forecasts are from planning to implement any artificial intelligence of customer relationship management (CRM) projects, the improvement forecast is of categories of activity , e.g. corporate marketing , digital commerce, and customer analytics. They are not estimates of ROI for the (AI) software. They rely on conservative estimates to which each of these entities might affect company revenue, expenses or productivity. They also rely on estimates of the penetration of software in customer relationship management activities . Net new jobs created are based on the ratio of new revenue to jobs required to support that revenue . They can assume that 50% of the net new revenue will support increases in labor and the rest will go for capital and other operating expenses that may replace jobs lost to automation.

In the future, some of the ways in micro economic benefits to any organizations. (AI) technology is expected to impact CRM activities include: Spending up sales cycles, improving lead generation and qualification solving customer support problems faster (raising service quality), helping companies improve brand campaigns and recognition, lowering costs of support calls when increasing resolution rates, lowering the cost of recruiting employees and partners, increasing revenue from optimized product marketing, optimizing price, distribution logistics and preventing loss through fraud detection. So, micro economic benefits view point, it seems that (AI) CRM technology can raise any companies economic benefits for care term.

Artificial intelligence enables machines or the in-build software to behave like human beings which allows these decisions and act. The advent of (AI) is leading , talking, making decisions and act. The advent of (AI) is leading to new technologies advances and transforming the economic and employment opportunities for humans in a positive way. (AI) related technologies can facilitate our live. For example, industrial robotics, robotic medical assistants, smart games, financial forecasting software, big data analysis, algorithms in health and bioinformatics, pilotless cargo places, drone ambulances and general purpose and workplace robots and others. (Disruptors technologies: Advances that will transform life, business and the global economy).

Artificial intelligence also known as computational intelligence is defined as " the human –like intelligence exhibited by machines or software. It is theorized that intelligence of humans can be described and intelligence machines or software can simulate it. These machines software can be reasonable , learn, perceive and process information, like human mind and thus facilitate human life. They can think and act for us. So, artificial intelligence is an interdisciplinary field of study including computer science, neuroscience, psychology, linguistics and philosophy.

However, (AI) research and developments have economically impacted many industries, such as robotics, telecommunications, computer applications , health, finance, heavy manufacturing, transportation, aviation, e-service and e-commerce, military , music and movie, toys and games entertainment etc. industries.

In fact, many ideas, systems and technologies have been developing in the world of (AI) technology. However, which are net called or considered (AI) products, rather which are mentioned with their specific names, such as smart graphics, machine learning, e-commerce etc. (i.e. this is called (AI) effect).

What is relationship between
(AI) and digital economy?

● How can (AI) technology influence digital economy?

Nowadays, (AI) related industrial applications will replace most human power in fields, including call centers, customer services and air cargo transportation. (AI) technologies also help weather forecasting based on repeated rainfall pattern (data) recognition, through robotics (i.e. floor cleaning, moving lawns etc.) transporting people and products with unmanned vehicles, sending space unmanned smart shuttles, developing robotic arms, predicting market values in stock exchanges by internet, making homes safer, helping elderly and disabled using robotic servants etc.

Among the (AI) related technologies , there are a few that significance for the impact on society and especially on digital economy . (AI) is particularly influential in machine learning. Such as robotics, transportation, finance, health and bioinformatics, e-commerce , e-games, big online data gathering and internet-of-things. For example, machine e-learning is based in bioinformatics and robots that can learn new skills for better caregiving in healthcare. What is machine e-learning? Machines can e-learn from e-data gathering, coming up generalizations and making decisions to act in certain ways from internet.

There are important applications , such as e-machine perception, electronic online natural language learning processing, online search engines, online bioinformatics, online brain –computer interface, online game playing, online robot locomotion, online advertising, online computations finances, online health monitoring, online DNA classification and decision making, online in chemistry –cheminformatics . So, online machine learning can positively impact productivity and it can enhance information and analytical system from (AI) online channel.

What is robotics? Robotics is one of the most strongly influenced fields in (AI). For example, heavy manufacturing industries, robots and used and man power is replaced for effectiveness, precision, and accuracy, especially in respective or dangerous tasks, including welding, assembling , picking and placing .

So, robots can acquire new skills or adapt the changing dynamic environment. Also, artificial intelligence can be applied in developing transportation. For example, automated vehicles, driver assistance systems , safety systems, collision avoidance systems and public transportation. Moreover, (AI) technology has proven to produce some of the best tools to predict stock market fluctuations from internet data gathering method. It's predictions are based on ever-evolving predictions algorithms and systems learn new models and make connections between historical data and new data to measure stock market trading more accurate from internet data gathering channel.

In health field, especially in health data processing , analysis, decision making support and medical diagnosis. So, online data can show which patients will need what treatment and what alternative drugs could be used more accurate from (AI) online data gathering method. Bioinformatics is an interdisciplinary field combining statistics, (AI) online technology can help in discovering data patterns and modeling through the application of machine learning, artificial neural networks and genetic algorithms. For example, further (AI) technology development of human genome project of online data sequences.

Online shopping can be facilitated by virtual assistants developed through (AI) technology and these assistants can offer the best advice. (AI) online purchase coming after every product image recommendations and personalization bring important revenue to shopping online sites, like Amazon . Smart computer graphics and games, artificial intelligence is useful in smarter computer, graphics, scene modeling , scene rendering processes in order to create, for example, effective human –robot interactions , online machine learning, online strategic games techniques etc. online computer related (AI) software.

So, online big data analysis and big data does have a critical need in the world of online intelligence machines and software in our future. In other words, (AI) offers online technology to enable online big data analysis to provide industrial organizations with valuable information for effective decision making in short time. For example, what IBM's Watson achieved: this machine used 200 million of structured and unstructured content with a special technology of hypothesis generation, massive evidence gathering, analysis and scoring from internet channel.

Finally, (AI) online technology another related internet invention (internet of things) (IOT) is the network of machines or objects connected through internet. These connected objects can sense their internal and external environment, communicate with each other, can send critical data and finally can make decisions to act or correct their environment from (AI) online technology. For example, factories can monitor and automatically change production processes, hospitals can monitor and regulate the health conditions of their patients , schools can collect data from facilities and cars can send data to car makers from (AI) online technology.

Partner predicts that (IOT) market will create about trillion amount value by 2020 year. Although machines collect big data from their environment, whether which gain an insight or learn from these online data largely depends on the (AI) online machine learning principals and (AI) online technology. In 2013, Mckinsey estimated that disruptive technologies closely related with potential economic impact in 2025 year between $7.1 to $13.1 trillion amount (automation of knowledge work, advanced robotics, autonomous or near-autonomous vehicles).

What is the relationship between (AI) and global digital economy development ?

● Could work activities in China be automated making in the nation with the world's largest automation potential? Can (AI) technology influence China economy? Could China workers be affected and jobs made up of routine work activities and predictable? Will programmable tasks be particularly impact to China employment market ? When impact on labor market is likely to be gradual at the aggregate level, it can be sudden and dramatic at the level of specific work activities, rending some job obsolete fairly. Overall (AI) technology will raise digital skills when reducing demand for medium incomer inequality for China workers. It seems (AI) technology's effect on productivity could be crucial to China's future economic growth as the population ages are increasing.

In China, some biggest technological companies driving significant investments in research and development. Moreover, China is one of the leading global (AI) technology development county. However, China will need to focus on building its innovation capacity. For example, United States and United Kingdom are currently producing more influential (AI) technological research. However, if China planed to achieve (AI) technology success, it's traditional industries will need to develop technical know-how —to and overcoming implementation costs prepare to develop (AI) . When (AI) technology is introduced into China society, China government needs to raise concerning ethical, legal, technological security etc. business questions. Also, surrounding issues include privacy, discrimination, legal liability and regulation. It aims to encourage overseas investors to choose to invest (AI) technological industry to raise GDP growth and manufacturing industries income growth for long term in China. If China encouraged overseas (AI) technology investment in its country. It is possible to influence China employment market to be changed. Because (AI) technology will impact to influence China people daily life. Due to (AI) technology is introduced to China society, many rich people will prefer to spend to buy any high (AI) technological products for entertainment or learning or machine man driving etc. daily necessity activities. Then it will raise GDP growth and will raise (AI) manufacturers or related-(AI) technological manufacturers profit. It is beneficial to China because it can become one high knowledgeable and (AI) technological economical society. But it will bring bad influences to raise unemployment chance for the low skillful labor. In labor economy aspect influence , how (AI) technology can influence China low skillful labor unemployment ratio raising. The raising low skill labor unemployment reason is because China low skillful human labors are argued or are replaced by (AI) technology creating new challenges to introduce to influence China society of simply human manufacturing job nature to be changed to be high (AI) technology manufacturing job nature in any China factories. Moreover, when (AI) technology introduction to China, it will cause other related social challenges in China. The varied (AI) related challenges, including the difficulty of creating safe and reliable hardware for sensing and affecting (transportation and education), the challenges of gaining public trust, a low resource comities and public safety and security, the challenges of overcoming fears or marginalizing humans in China employment and workplace and the risk of diminishing interpersonal trust because the low skillful labors won't believe any China employers will give chance to employ them , due to (AI) technology will replace their skills and man manufacturing of productivity is much less to compare to (AI) technology manufacturing method.

How does (AI) technology influence
the future of employment change?
Are future nature of jobs changed to computerization from (AI) technology? Where are the probability of computing occupations from (AI) technology influence? What is expected impacts of future computing on labor market from (AI) technology influence? John Maynard Keynes's frequently cited prediction of widespread technological unemployment " du to our discovery of means of economic the use of labor outrunning the pace of which we can find new used of labor" (Keynes, 1933, p.3).
In the future, (AI) technology will impact some nature of occupations to change computing. This chance will also influence some countries' economic change. For example, some factory human labors hand routine manufacturing tasks will be changed to computerization of routine manufacturing tasks by (AI) technological machine men hand manufacturing method. it will cause a structured shift in the labor market, with workers reallocating their labor supply from middle-income

manufacturing to low-income service occupations.

Arguably, this is because the manual tasks of service occupations are less computerization, as who require a higher degree of flexibility and physical adaptability. So, (AI) technology will influence the human hand labor skillful occupation nature of task cheaper , such as vehicle manufacturing , ship manufacturing, computer manufacturing, steel manufacturing, television, radio etc. home electronic products of heavy machine industry change. Due to (AI) technology machine man will be proper to be used to manufacturing these electronic products when the (AI) technology innovation can develop to the mature stage. Then, any countries manufacturers will choose to use (AI) technology machine man, instead of human hand production.

Supposing the future prices of computing are fallen, seriously, problem solving skills are becoming relatively productive, explaining the substantial employment growth in manufacturing occupations, involving cognitive tasks where skilled labor has a comparative advantage, as well as the increase education needs for (AI) technology computing of machine man subject study.

Prediction of education needs for (AI) technology student numbers will increase, due to manufacturing industry needs many (AI) technology students in future employment market. Another (AI) technology influence if the future (AI) technological innovation, e.g. machine man manufacturing or machine man service industries will both increase demand, then with more sophistic software technologies will be disrupted labor markets by marketing workers redundant.

For publishing industry, what is striking about the case in paper book publishing industry will be unpopular? Due to the electronic book publishing industry will be popular, e.g. Amazon publish . (AI) technology can influence paper book manufacturing method which is replaced by machine man electronic book manufacturing method as well as it will cause the computerization is no longer confined to routine manufacturing tasks. Due to (AI) machine man manufacturing technology will be proper to be used to manufacture any products in short time efficiently and effectively , e.g. electronic book products. In the future, if it is fact to occur this case, such as (AI) technological machine man manufacturing method will be adopted (applied) to manufacture electronic books or any products in possible. (AI) technology will cause many manufacturing workers are unemployed. It is beneficial to employers, who can reduce to spend much wages expenditure to employ manufacturing workers, but it will cause many manufacturing workers loss jobs and reduce income to support whose families lives. It will cause social challenges, e.g. increasing stealing crimes if the manufacturing workers had not other skills to find other jobs to do easily. So, manufacturers need to concern over technological unemployment which will be hardly future phenomenon if who decided to dismiss all manufacturing workers, due to (AI) technology machine men replace to them.

If (AI) technology can be innovated to produce any kinds of machine man to serve any service or manufacturing industries successfully. Then, it will bring these questions: Can future that workers be influenced to be automation employment and productivity by (AI) technology influence? Does it impact to influence the (AI) technology countries' productivity and growth and natural resources development and labor markets and evolution of global financial markets and economic impact of technology and innovation and urbanization etc. issues? How will automation transform the workplace? What will be the implication for employment? What is likely to be its impact both on productivity in the global economy and on employment?

In fact, automatic of activities can enable businesses to improve performance by reducing errors chance and improving quality and speed, and same cases achieving outcomes that go beyond human capabilities. Some economists indicate (AI) technology would give a needed boost to economic growth and prosperity have of the working age population in many countries. Based on the scenario modeling, they estimate automation could raise productivity growth globally by 0.8 to 1.4 % annually. They also indicated that almost half the activities people are almost $1.6 trillion in wages to do in the global economy have the potential to be automated adapting current demonstrates technology, according to their analysis of more than 2,000 work activities across 800 occupations. When less than 5% of all occupations can be automated entirely using demonstrated technology, about 60% of all occupations have at least 30% of worker made activities, that would be automated. More occupation will change to be automated. They also indicated for business performance benefits of automation are relatively clear, but the issues are more complicated by policy making to attract foreign investors. Beyond technical feasibility, the cost of technology, competition labor will include skills and supply and demand dynamics, performance benefits and beyond labor cost savings and social and regulatory acceptance will affect the automation. Their predictions suggest that half of today work activities could be automated by 2055 year, but this could happen 10 to 20 years earlier or latter depending on the various factors in addition to their wider economic condition.

Some scientists suggest (AI) technology is finally starting to deliver real-life business benefits. Computer power is growing significantly , algorithms are becoming more sophisticated and perhaps most important of all, the world is generating vast quantities of the fuel that powers (AI) technology data billions of gigabytes of it every day. Also, online firms are digital natives, such as Google online search service company is investing on (AI) technology. For new though most of the news if coming from the suppliers of (AI) technologies. And many new users are only in the experimental phase. Few products are on the market or are likely to arrive these soon to drive immediate and widespread adoption. As a result, analysts believe (AI) technology's potential will give true economic benefit in the future. (AI) industry will introduce to suppliers and users to raise economic potential of (AI) technology.

In the future, (AI) technology systems can solve business problems. Some scientists categorized those into five technology systems that are key areas of (AI) technology development: robotics and autonomous vehicles, computer vision language virtual agents and machine learning , which is based on algorithms that learn from data without replying on rules-based programming in order to draw conclusions or direct an action.

Such as computer vision and language includes natural language processing, analytics, speech recognition technology, some are about learning from information, such as about machine learning and others are related to acting on information, such as robotics, autonomous vehicles and virtual agents, which are computer programs that can converse with humans. Machine learning and a subfield called deep learning are artificial intelligence applications.

- Can artificial intelligence impact

global economy growth?

Artificial intelligence (AI) is a term first defined in 1956 year. It is a branch of computer science that aims to create intelligent machines that work and react like humans. In contrast today, 60 years later, (AI) is characterized by a number of applications, including computers playing games against humans and understanding human languages, virtual personal assistants, and robotics which involve computers seeing , hearing and reacting to sensory stimuli. In the future, technologists predict for (AI) technology ranging from (AI) being used as a tool to aid relatively simple processes for robots with human like mental capabilities, who expect (AI) technology can emulate human performance by learning, coming to mind its own conclusions, understanding complex content, engaging in dialog with people, enhancing human cognitive performance or replacing humans in executing both routine and non-routine tasks. In existing industry, (AI) technology is used , such as targeted advertising and virtual used personal assistant as well as the (AI) technology that my exist in the future, such as robots with human vehicle processing capabilities.

The range of (AI) technology's progress in the future will determine the economic impact future of (AI) technology on the global economy with more limited advances and applications (i.e. weak (AI) only) corresponding to more limited economic impacts and more substantial progress, i.e. strong (AI) technology is corresponding to more significant economic impact.

(AI) technology learning that automates analytical model, including predicting cause-and-effect relationship from biological data, identifying new drugs, self-driving cars and protecting against fraud etc. functions. Also (AI) learning can improve natural language processing that allows computers to continue to better analyze, understand and generate language to interface with human using the natural human language, virtual personal assistant, helps users by providing scheduling appointment, reminds organizing personal finance and finding providers of various services, machine vision allows (AI) machine man to identify object, scenes and activities in detect pedestrians and bicyclists.

We expect the economic effects of (AI) technology to include both direct GDP growth from sectors that develop or manufacture (AI) technology and indirect GDP growth through increased productivity in existing sectors that employ some from of (AI) technology. If (AI) producing sectors could grow, then it could lead to increase revenues and employment of (AI) technological professionals within these existing firms as well as the potential creation of entirely new economic activities to any countries' societies productivity improvement in existing sectors could be realized through faster and move efficient processes and decision making as well as increased (AI) technological knowledge and access to information available in societies easily.

In the future, if (AI) technology is an increasingly critical component of more products, it will become an integral part of necessary products of many people's lives. The extent of (AI)'s economy effort is also likely to vary from region to region, thought variation may be more dependent on the predominate economic activity of a region and the (AI) ability can influence economic activity, rather then the economic or developmental status of the regions. (AI) technology can move accessibility and can use source development to do international business between one country and another country.

So (AI) technology has the potential to give benefits to different income chooses and to bring significant gains to both developed and developing countries. For agricultural technology, (AI) has the potential to optimize food production around the world by analyzing agricultural regions and identifying what is necessary to improve crop yield. In total, (AI) technology gives greater economic impact to any countries agricultural regions if which implemented (AI) technology to grow crop , fruit etc. food production in the farms.

Investment in (AI) technology is such as capital investment to any countries' public or private enterprises. So, it will have large economic impact to the future . If the (AI) technology is reasonable invested to the different needs aspect by the public or private enterprises in the country. Then, it will have good economic impact to the country in the future. However, when (AI) technology is likely to affect both the productivity and employment components of economic growth in many sectors. Significant public debate has focused on projections of (AI)'s effect on the labor force. However, for instance, some researchers have argued that the rise of (AI) technology and automation will led to significant unemployment as capital is substituted for the low skillful labor. So, they point to the concern that the increasing sophistication of (AI) technology may balance skilled and semi-skilled workers and the reduce the size of the middle class. However, this is not a new argument, due to (AI) technology negatively affecting the labor force and leading to mass unemployment. Because the (AI) technology is the substitution of machinery for human labor. Although, employment in certain industries, has been reduced in the past due to technological advancement. For long term, the labor market has adapted to the introduction of new technology, giving rise to new jobs in new areas. (AI) technology may also be accomplished without a reduction to total employment in the long-term to some Asia countries, such as Hong Kong and Japan. Because Hong Kong and Japan many low skilled labor, e.g.

security, cleaner who complaint that employers need them to work long time hours. (abnormal working hours) e.g. one day 12 to 15 working hour per day. Hence, if (AI) machine means invention technology success. Security or cleaning job can be worked from (AI) machine man in some hours every day in order to reduce the long time working hours cleaners or security workers, e.g. one (AI) machine man works 4 hours for cleaning or security job, one day as well as another cleaner or security labor only needs to work 8 hours one day. So total security or cleaning employers can employ 12 hours machine cleaners or security workers and human cleaners or security workers in one day. For long term benefit, Hong Kong or Japan every security or cleaning worker does not need to work 12 hours minimum working hours one day. They won't feel tried and bore and without private with whose families, so who will accept to do these cleaning or security jobs, even they can raise work efficient and performance when who feel happy and health.

So, (AI) technology of machine man invention can raise low skillful labor efficiency and it can help them to avoid abnormal working hours demand in some busy work life countries, such as Hong Kong and Japan. Before, one Japan female labor feel unhappy to work, due to who often needs to work abnormal working hours for her employer and who has less sleeping and without any private time to enjoy her life with her families every day. So this abnormal working hours factor causes her to do commit suicide behavior, then she is die unlucky. So (AI) technology of machine man invention ought avoid abnormal working hours demand for employer in any countries in the future.

The most important occurrence to any employers, some researchers had attempted to do one experiment to find that private research and development , venture capital and public research and development investment all have strong net effect or economic growth with venture capital funding further having the strongest such effect from (AI) technology. The researchers hypothesize the venture capital investment contributes to economic growth through (AI) technology innovation and by the capacity of an economy to use existing (AI) technology knowledge to increase productivity. They predict the impacts of venture capital, business-research and development and public research and development can raise multi factor productivity from (AI) technology introduction.

Can (AI) technology influence the economic development to developing countries? The developing regions of the world contain most of natural resources. If one day, (AI) technology has invent one kind of machine man which can assist any gas or oil workers to seek any new oil/gas natural resource locations easily. I believe that (AI) technology can help these natural resource exploitation countries will gain economic benefit more easily. So, (AI) driven technology can be used to change to create any new opportunities to address poor management or resources and improve human well being, such as Africa Latin America and India can use (AI) technology machine man to seek any oil/gas natural resource countries exploitation activities to attempt to gain much economic benefits.

● Why will (AI) technology grow economic
development ?
Nowadays, increases in capital and labor are no longer driving the levels of economic growth, such as (AI) technology. The ability of increase in capital investment and in labor of traditional drivers of production, have no longer to be enjoyed in most developed economies ,e.g. developed country, US, UK . However, artificial intelligence has the potential to overcome the physical limitation of capital and labor to avoid missing out on this opportunity. So, policy makers and business leaders must prepare for and work toward a future with artificial intelligence. They must do with the idea that (AI) is another simply method to enhance productivity method . Rather they must see (AI) as the tool that can transform thinking about how growth is created.

Economists have always thought of new technologies are as driving growth their ability to enhancing. It can replace labor and capital factor of production. So, it brings this question: What is the factor of production (AI) technology characteristics. They key factor is to see (AI) technology as a capital-labor .

(AI) can replicate labor activities at much greater scale and speed, and to even perform some tasks began the capabilities of human. For example, by using virtual assistants , 1000 legal documents can be reviewed in a matter of days instead of taking three people six moths to complete. Some (AI) technology may be one kind of factor of production in the future. For another example, people will work in workplace digitalization environment. So, in the future, working environment and information management are automated. Such as Konica camera sale company will use workplace digitalization. So , (AI) technology can provide workplace digitalization in order to raise productivity efficiency. (AI) technology will be one kind of production which is replaced by workplace digitalization and it will grow any organization productivity efficiently. Then, (AI) technology will assist overall social economy growth , due to productivity is raised and products can be produced in short time to prepare to sell in consumption market. So, time will be shortened to increase GDP growth fast for the development of (AI) technology countries.

● How can (AI) technology impact to global
economic and social and psychological
changes?

What will be the development of (AI) technology and predictions concerning the future evolution? The computers and robots will develop conscious, intelligent and minds into humans, enhancing psychological and behavioral abilities and allowing for direct communication with (AI) minds. (AI) technology will be impacted human life by (AI) technology information communicative and environmental influence. A " world brain" and " world mind", this psychological system will be enhanced and enriched the capacities of both individual and collective cognition by (AI) technology of service industries.

(AI) technology with influence these human needs of service industries changes, such as , biological science, finance, entertainment, business, biological science, transportation, communication military etc. The personal computer evolution, the internet and the world wide web which exploded on the scene, linking business, homes, schools, social organizations which were a completely unpredicted phenomenon to influence human life. Kurzweil (1999) predicts that by 2029 year, most human communication will be with machines. According to Person, by 2100 year, there will be human machine convergence.

How can (AI) technology influence environmental protection to make benefits to farming economic growth? (AI) technology can be applied to predict how to solve environmental pollution challenge to avoid to damage any crop or vegetable or rice or fruit etc. food growth. Because environmental experts can gather global environmental pollution data from an environmental database to build a perform a systematic analysis from (AI) technology. The first step is this broad analysis can include understanding, statistical and data gathering techniques to obtain the relevant data, the correlation among the variables involved, and a list of possible models. The next step is to select a set of methods and models that cover all kinds of knowledge and functionalities needed for the decision making process. Once the models are selected, they must be fully implemented by means of machine learning , data mining, statistical or numerical technique. After that, those models must be integrated to build the whole EDSS. The EDSS must be tested to check its performance, accuracy, usefulness and reliability, both from the user's and (AI) technology/computer scientist's point of view. If these is any wrong feature in any development stage, such as model's integration, models' implementation, selection of models, database, problem analysis etc. the developers must come back in the update th required components. When the evaluation phase is all right, the EDSS is ready to be applied to the environment. The great contribution of artificial intelligence to EDSS the integration of several methods complementing the classical statistical models/simulation , statistical analysis, linear models, etc. and numerical models (control algorithms, optimization techniques etc.) .

This cooperation makes the resulting systems more reliable and powerful in coping with real world environment systems. Date interpretation has been a principal area of research in (AI) technology since the very beginning. The most demanding problem in the environmental assessment context. Knowledge representation permits the definition of the different types of data that the existing methods adapt to the process. There is also a lot of work to clean, repair and transform the huge available quantities of raw data. Apart from this, the availability of meta-information or background knowledge is required to guide the process. Data mining is multi-disciplinary: It covers expert systems, data based technology, statistics, data visualization and unsupervised machine learning. These techniques operate at the level of data and background information, where numerous and often incompatible new commensurate pieces of information from disparate sources have to be brought together (K, Fedra, 1994).

So, it seems that in the future, (AI) technology with the increasing maturity in particular those related to knowledge and engineering, new dimensions can be assisted to users in environmental decision making are available. For example, many environmental systems are characterized both by incomplete models and by limited data. Hence, in the future, (AI) technology will be applied to predict climate change to reduce crop or fruit etc. food agriculture challenge by climate change bad influence.

● Will (AI) technology influence digital economy change to manufacturing industry ?

To understand how the manufacturing business must adapt to prosper in the technology, we need to understand how (AI) technology will change us to shape our daily habits to satisfy our expectation of products to how we shop and even the immediate of the entire process. For example, taxi services are in the crosshairs as on demand transportation services like, available of the touch of a smart phone button expand. In fact, Yellow lab, US country , san Francisco city's largest taxi company is filing for bankruptcy as the industry starts to change faster than almost anyone expected. However, at this point, its more than an app that is changing, some our taxi passengers renting taxi transportation to catch consumption behavior.

(AI) technology will influence digital economy for taxi passenger's individual customer experience, offering a growing renting taxi to catch of service and feedback opportunities when any one taxi passenger who chooses to use mobile phone app online tool to prepaid to rent any taxi more easily.

Also in the long term, (AI) technology can influence vehicles drive themselves of behavior. Already, companies like Google and GM are working on projects to bring fleets of autonomous vehicles to cities at the path of a button.

Moreover, this on-demand service model is beginning to appear across a much broader range of markets. For example , Amazon company is investing in its own fleet of trucks, planes and even drone at the same time as it pushes for same-day delivery of products. As some point, vehicles will be autonomous too. So, it seems that (AI) technique will influence any transportations choose to use digital autonomous driving technology in the future . For Amazon company case, it is not stopping of logistics. It is also aiming to automatically manage the supply of consumer home products with its recently launched Amazon replenishment service, Dash. Dash is a digital service that enables that connected derive to automatically order physical products from Amazon when supplies are running low. So, it seems (AI) technology will be applied to logistic function by digital technology method introduction in the future.

Hence autonomous vehicles will optimize industry supply chains and logistics operations through increased efficiency and flexibility. In fact, fully automated and lean supply chains will keep reduce load sizes and inventory by leveraging smart distribution technologies and smaller autonomous vehicles by machine man assistance. If Amazon continues to grow market share for online sales by reducing effort required by the consumer to place an order, when also contributing the almost immediate delivery of products to the doorstep. So, it will further fuel the trend toward on-demand derive. As Amazon company fuels the on-demand economy, consumers will expect immediacy in more parts of the digital economy. On top of speed, consumers increasing expect more personalization options.

So, (AI) technology will influence digital manufacturing, such as Amazon publishing to monitor every aspect of every process in real -time and communicating to self-optimized deep learning robotics, new methods of high volume and high customization will become possible. Then, as products merge into product platforms and even services, manufacturers have the opportunity to provide components and platforms used by smaller players. So, (AI) technology will influence manufacturing industry to choose automated SMI lines, robots installed, automation engineers.

Another future (AI) technology development can be applied to space science aspect, such as Automation engineering space in manufacturing process to achieve digital manufacturing benefits to any businesses in the future. Such as reducing cost, shortening manufacturing time, raising efficiency, shortening delivery products to client individual time. How can artificial intelligence give the need and advanced fast and evaluation methods benefits for space exploration? When US NASA (space exploration organization) achieves any space exploration missions, it will answer this question:

When is it useful to have a machine use (AI) technology to achieve a decision? After all, after millions of years of space exploration and rough 10,000 years of civilization, humans are usually quite good at making decisions in complex uncertain environments. Through, Johns Hoplains University's Applied Physical Lab. Research in (AI) technology enabled systems, which has identified three general use cases for (AI) technology to explore space mission:

First, for some tasks (AI) technology is more cost effectiveness than human. Second, (AI) technology is better suited than humans at solving some, but not all problems. Third, (AI) technology allows NASA organization's space exploration mission to develop machines that ate capable of responding faster than when a human is in the decision loop (D. Scheidt, 2012, A. Castano et. al. 2008).

So, the use of (AI) technology to enable science by observing the pace of rapidly evolving phenomena was demonstrated. It is more effectively coordinating and (AI) technology utilizing to earn economic benefits to use for space exploration mission.

However, (AI) technology also have current risk for space exploration. Today (AI) technology is immature and requires further development to reach its potential. For instance, the (AI) technology algorithms that detected the dust derive could not have identified whether the Martain weather represented a threat to the cover. Also it can not yet use instrument input to determine what, where and how to autonomously make the next space science measurement. An equally important factor limiting (AI)'s deployment is that lacks the methodology and technology to effectively test (AI) technology. So, the challenge will testing (AI) enabled system is how (AI) performance can be measured. It would be NASA organization's difficulty to find (AI) technology to develop to carry on researching any space exploration missions in the future. However, (AI) technology will be a good economic benefit choice for space exploration mission in the future.

● What is artificial intelligence potential benefits and ethical considerations?

The ability of (AI) technology systems to transform vast amounts of complex information into insight has the potential to help solve manufacturing or service challenges for human needs. However, to reap the societal benefits of (AI) systems, humans will need to trust then and make sure that which follow the same ethical principles, moral values, professional codes and social norms that we humans would follow in the same scenario, research and educational efforts as well as carefully designed regulation in order to achieve the most effort of economic benefits goals. For example, international business machines corporation (IBM) is actively engaged both competitors , in global discussions about how to make (AI) ethical and as beneficial as possible for people as social economic benefits.

(AI) is usually defined as the " capability of a computer program to perform tasks or reasoning processes " that human usually associate to intelligence in a human being. Often, it has to do with the ability to make a good decision, even when there is uncertainty, too much information to handle. As an example, play chess or complex card games of entertainment activities is believed to need some form of intelligence in a human being, as well as choosing the best medical facilities in a difficult medical case, or creating something new, such as mathematical theorem or even some form of act, or even driving automatic machine man (self driving vehicle) replacing human driving in the middle of a crowded city.

(AI) needs depends on what we consider being intelligence in the behavior of a human being act a certain point in time. If human belief about human intelligence changes and we don't believe any longer that a certain task requires intelligence, then a computer program performing that task is no longer part of (AI), it becomes just another boring computer program. So, it means that (AI) technology will replace some old computer programs, if human can invent new generation of (AI) software for any functions or activities to satisfy human needs.

As IBM, it argues intelligence. This means that we aim to build systems that enhance and scale human expertise and skills rather than replacing them. We therefore focus on practical applications of (AI) capabilities that assist people in performing well-defined tasks of needs by exploiting and wide range of (AI)-based services. We also use the term " cognitive computing" it is mean a comprehensive net of capabilities based on technology. It comprises the fields of machine learning, reasoning and decision technologies, language, speech and vision recognition and processing technologies, high performance and high efficient functions for any industries or individual consumers needs. For example, robotics, which are usually very good at doing what which are supposed to in any environment, much have public shopping center, factory etc. places which need simply services from the robot (machine man), such as cleans the floor of our houses to the robot that can work together with humans in production chains, passing through the warehouse, robots can take care of the tasks of an entire warehouse and the companion robots like Nao, Pepper, Aibo and Giraff, who can entertain use, talk to use and help elderly people to stay connected to their friends, relatives and doctors.

Google company is building automatic machine (self-driving cars) and has acquired more than 10 robotics companies. Facebook had opened whole new research facility only on (AI) research. Apply computer has developed Siri. Microsoft computer company has built a similar personalized assistant. Google has Deep mind, a UK company whose long term aim is to build general (AI) and has already great potential to win game to the world champion and IBM is investing a huge amount of resources in applying its Watson cognitive computing system to the medical domains to finance and to personalized education. In Europe, IBM is establishing new centers in Munich and Milan focused in the application of cognitive computer capabilities to the internet of things and healthcare respectively.

For example, automatic machine man (self-driving cars) are all about (AI), which used to be able to see what happens in the street (signals ,lanes, other cars, pedestrians, traffic lights, which need to able predict what other cars and pedestrians will do, and who need to be able to cope with unforeseen situations. Since, most car accidents are due to human fault, it is estimated that the adoption of self-driving cars will save about half of the lives that are usually last in car accidents.

IBM Watson company has to understand spoken language, make sense of massive amount to text , respond correctly to questions in many categories, as well as assess its own confidence in responding to such questions. In the future, (AI) technology can own question/answering capabilities that would be very useful, for example, in assisting a doctor when trying to some to the correct diagnosis for a patient and to propose the best therapy .

Intelligent machines can also rely on huge amounts of data to be used to learn how to make better decisions. This data comes from all of us over the years Facebook users have uploaded more than 250 billion pictures and every day who upload about 350 million more. Every second, we submit 40,000 google search queries. So, (AI) technology will be connected through the web from appliances to traffic lights from cars to watches. Other tasks that are very easy for humans are physical and manipulation tasks, such as walking , running, picking up an object to make its shape and location, restricted environment. But (AI) machine man technology still not able to have the general physical and manipulation capabilities even of a 6 year old.

So, it brings this question: Why do (AI) scientists need to concern ethics? Because (AI) technology is complex, information into insight has the potential to reveal long held secrets and help solve some of the world's most difficult problems. (AI) systems can potentially be used to help discover insights to treat disease, predict the whether, and manage the global economy. So, ethic issues is important to and (AI) scientists . If any one new (AI) technology research investigation could success, it will be a secret to and the (AI) scientists can not permit to their loyalty to any competitors to damage the fair (AI) technology products trading market. The country (countries) (AI) technology scientists need to concern ethic issues, who need to keep secrets for their countries economic or/and social benefits. This is moral issues to any countries/country loyalty is whose countries intangible assets. They can not sell (AI) loyalty to any their countries to assist whose economic benefits immorally.

● How can (AI) technology influence to global health care economy development?

According to (AI) lecturer analysis, when combined key clinical health (AI) application can potentially create $150 billion in annual savings for the US healthcare economy by 2026 year. (AI) technology is re-winning modern conception of healthcare delivery. It enables machines to sense, comprehend, act and learn. So which can perform administrative and clinical healthcare functions (Accenture, 2017).

It will help health care service organizations to reduce health care cost, will improve and raise service quality and access. So, (AI) health market size will be predicted growth. (AI) applications in health care include robot-assisted surgery, virtual nursing assistant, administrative workflow assistant, fraud detection, error reduction connected machines, clinical trial participant identifier, preliminary diagnosis, automated image diagnosis and cybersecurity.

What kind of benefits (AI) technology can contribute to healthcare service? (AI) technology can deliver what many health care organizations need, such as financial and operational of labor costs, digital expectations from patient consumers how to use (AI) technology to solve interoperability challenges in any healthcare organizations. Also (AI) technology can be applied to wellness an d lifestyle management, diagnostics, delivers financially but also way of organizational and workflow improvement. So, (AI) technology will be continue to become most prevalent and adoption to healthcare organizations , which must need to enhance structure to be position to take full advantages of new (AI) technological capabilities. (AI) technology can change the nature of work and employment is rapidly changing to make the best use of both humans and (AI) talent in healthcare industry in the future. For example, (AI) technology offers a way to fill in gaps and the rising labor shortage in healthcare. According to Accenture analysis, the physicians shortage is increasing. However, (AI) technology will manufacture healthcare machine men to replace physicians in future one day(2017). Hence, (AI) technology will be invented to raise health care service staffs work efficiency and performance in any hospitals or clinics in the future.

In conclusion, (AI) technology will raise efficiency for any service or manufacturing industries in the future, although, it is possible that it will also rise low skillful workers unemployment numbers. But, the most important influence to human technological innovation will be risen and it will influence human life will be changed to be better, e.g. self drive cars, health care physician machine men, machine man cleaners etc. intelligent machine men will be manufactured to serve for our daily life. Furthermore, (AI) technological products will influence countries trading, some low technological development countries manufacturing businessmen can choose to buy any (AI) products to raise whose productivity and efficiency and reducing cost to achieve economic cost saving result. Also, GDP of trading growth income will increase to the (AI) products sale countries. Hence, it will be beneficial to economic development to both developed and developing countries both in the future as well as (AI) scientists time and money spending will be valued to continue to invest (AI) technology development for human life and economy benefits for long term.

In consequent, when (AI) can spend more time to attempt to do human's any tasks every day, then it have possible to raise productivity and efficiency better than human.

Reference

A. Castano et. al. " Automatic detection of dust devils and clouds at Mars" Machine vision and applications, Oct. 2008, vol. 19, no 5-6, pp. 467-482.

Accenture, " Why artificial intelligence is the future of growth"(2017) <http://www.accenture.com/us-en/insight-a   rtificial-intelligence-future-growth>.

D. Schedidt , Unmanned Air Vehicle Command And Control, Handbook Of Unmanned Air Vehicles, Springer-Verlag, 2014. Facebook (AI) Research Available at https://research.facebook.com/ai, research at google, machine intelligence available at

http://research.google.com/pubs/machineintellige nce.html; micro soft research-machine learning and artificial intelligence available at http://research.microsoft.com/en-us/research-   areas/machine-learning-ai.aspx.

K, Fedra , "GIS and environmental modelling" in environmental modelling with GIS, edited by M.F. Goodchild.B.O. Parks and L.T. Steyaert, Oxford University press, pp. 35-50, 1994.

Keynes, J.M. (1933). Economic possibilities for our grandchildren (1930). Essays in persuasion, pp.358-73.

Mckinsey & Company (2013, May). Disruptive technologies: Advices that will transform life, business and the global economy , USA.

Ray Kurzweil , The age of spiritual machines (1999) is cited numerously through this chapter: Kurzweilai.net http://www.kurzweilai.net

Rich, Elaine & Knight, Kevin, Artificial Intelligence Second Edition, 1991, New York; Mc-Graw-Hill.

CHAPTER V

How Robots bring positive impact to raise organizational productive change

What is (AI) consumer behavioral prediction tool? How any why will (AI) tool assist manufactures to attempt to predict consumer behavior before and after consumption occurrence? First, I shall indicate how to apply (AI) tool to predict vehicle product consumer behavior case example.

Nowadays, many vehicle manufacturers hope their vehicles can attract to vehicle buyers to choose to buy their vehicles. However, there are many different brands of vehicles to provide to them to choose, so the vehicle market competition is very serious.

How to judge their different kinds of vehicle price which is reasonable acceptance to attract vehicle buyers to choose to buy the brand of vehicle manufacturers' any kinds of vehicles, e.g. fast speed sport style vehicles, comfortable and slow speed common cars, for four passengers common small size or more than four passengers common large car size? How to evaluate the vehicle prices issue is important factor to influence vehicle buyers' choices. Either if the brand of vehicle price is too high to compare brands, it will influence many vehicle buyers choose to buy other brands' vehicles or if the brand of vehicle price is too low, it will influence vehicle buyers feel this brand's vehicle's quality is worse to compare to other vehicle brands' similar vehicle products.

Thus, if the brand of vehicle manufacturers can predict how to design vehicles which can attract many vehicle buyers to choose to buy whose any vehicle products. What are future vehicle buyers' favorable vehicle styles? Then, the vehicle manufacturer can concentrate on manufacturing the kind style of vehicle products to sell already. It will reduce its vehicle manufacturing investment risk.

How to apply (AI) tools to predict vehicle buyers' behavioral consumption model? Whether artificial intelligent tools can predict automotive buyers' behavioral consumption model and predict future trend. In fact, automotive brands and dealerships are facing an increasingly competition when attempting to manually gathering the vast quantities of data required to create customer focused programs that increase retention, ultimately new sales and service automotive business. Building a based on that client's intrinsic needs and interests to any kinds of automotive vehicles at any given time. This is especially true in the automotive industry where the time span between purchases is measured in years. Because vehicle buyers would not like often to change their old vehicle to another new one. So, their decisions to buying another new vehicle, the time is usually after one year, even longer time. Hence, it seems any vehicles won't be frequent consumption products to the owned at least one vehicle family consumers (vehicle buyers).

Hence, how to predict vehicle consumers' taste or preferable which styles of vehicle choices issues is very important. If the vehicle manufacturers can not manufacture any attractive vehicles to sell easily in this year. Then, it will lose time, money in this year because it won't know when the owned least one vehicle users or non-owned any vehicle users who will decide to buy one new vehicle or change another new vehicle ensure. The different brand vehicle dealers will possible wait more than one year to attract them to buy their vehicles if their styles are not attractive to compare other brands of vehicle competitors.

However, artificial intelligence and machine learning can help any vehicle manufacturers to find solution to solve patterns in highly to solve patterns in highly complex data-sets that are beyond the capability of a human brain, and then building and automatically acting on the customer insights it generates.

Given the automotive customer need for individualized communications, this technology is positioned to become a critical component of any successful vehicle retailer's domestic or/and overseas vehicle markets. How can vehicle manufacturers and retailers use (AI) to enhance their vehicle marketing campaigns? How will (AI) affect their vehicle sale marketing strategy? What criteria would they use when selecting on (AI) solution?

Vehicle consumers today are able to quickly access different brands of vehicle information, research vehicle products and reviews, negotiate prices and compare one vehicle brand or retailer to another resulting of the brands of vehicle customers. At the same time, the rise of " big -data mining", wearable devices that track user's every move and preference and greater contextualization in advertising and social media has resulted in consumer expectations of individualized. Thus, it seems that (AI) tools can be used to gather " big-data" and then they can make human's mind to analyze how to design kinds of vehicles to satisfy vehicle buyers' needs.

As automotive vehicle marketers can apply (AI) tools to achieve messaging strategies to meet the needs of this new generation of informed vehicle consumers, using data from a variety of sources to move from a variety of sources to move from mass- messaging to more personalized messages aimed at particular vehicle buyer segments, e.g. fast speed sport vehicle buyer segment, slow speed comfortable small size or large size of buyer segment. However, when 90% of vehicle marketers believe having a single vehicle buyer view is important, only 6% have achieved it.

However, one of the main issues vehicle marketers facing is the lack of capacity to efficiently sift through and analyze the massive vehicle buyer amounts of data required to create vehicle buyer individualized vehicle customer experiences easily. This is especially difficult for automotive dealers, the long periods between purchase cycles, and the highly considered nature of the vehicle purchase means that each vehicle dealer needs to not only track a large number of potential vehicle customers for an extremely long period of time, but each of those vehicle customers will generate a huge amount of different kinds of vehicle behavioral consumption data as they research their next vehicle purchase. However, by choosing the right (AI) technological tools and programs , vehicle dealers can solve this big data gathering challenge into a major advantage.

For Forrester vehicle brand example, vehicle consumers have more power over the Forrester vehicle brand's reputation than ever before. Mayne, L. (2014) indicated that Forrester calls this new (AI) tools is the " age of the vehicle customer", a 20 year business cycle in which the most successful vehicle enterprises will reinvent themselves to systematically understand and serve increasingly powerful vehicle consumers. To win in this new age, Forrester declares companies must become vehicle customer obsessed and the only sustainable competitive advantage is knowledge and engagement with customers, such as (AI) gathering data knowledge.

Thus, the biggest challenge vehicle businesses currently face is not the collection of a large quantity of vehicle consumer data, but what to do with that data once they have it. Even at a large vehicle data research firm, the data sets are often too big for a single analyze, or even a team of analysts to sort through and draw conclusion from. However, enter artificial intelligence and machine learning , an efficient technology solution that can continuously find patterns in highly complex data sets that are way beyond the capacity of a human brain and then automatic drive action based on the customer insights is generated.

What is (AI) machine learning tool? Machine learning is a type of (AI) that learns from data and is not explicitly program. Think Amazon, face book. Machine learning serves up relevant content based on an individual vehicle purchase behavior and experiences. More simply, machine learning is a computer program that can learn relationships between data, subject those learnings to errors functions, and then learn from its errors. The program in effect, trains itself.

Lee, T. (2016) explained that "Thus, (AI) tools can learn deep a more advanced branch of machine learning inspired by how our brain's nervous function, has also been found to be especial effective in identifying patterns from data."

When this way sound is complicated from a vehicle dealer perspective, the implementation of a marketing program driven by artificial intelligence can take care of these tasks in an automatic vehicle fashion with little to no manual intervention required from the staff at time vehicle stores.

In practice at a vehicle dealership, the program will continue track vehicle customer behavior online, merging that data with any offline source (like CRM or DMS data) and then analyze this aggregated vehicle buyer data set to predict what vehicle customer may be shopping for and what information they might like to relevance from different kinds style of vehicle design photos.

1.1 Why can (AI) be applied to predict consumer behaviors?

Artificial intelligence refers to complex in vehicle market, machine learning that posses the same characteristics of human intelligence and that have all our sense, all our reason and think just like human do. Besides, machine learning is the practice of using algorithms to collect and examine data, learn from it, and then make a determination or prediction about something in the world.

The machine is " trained" using large amounts of data and algorithms that give it the ability to learn how to automatically perform a task with increasing accuracy. Otherwise, deep learning is primarily based on artificial neural networks inspired by our understanding of the biology of human's brains.

Deep learning breaks down tasks in ways that enables machines to assist us with increasingly complex tasks, driverless cars, better preventive healthcare and more accurate product recommendation (including vehicle recommendations). So, such as why (AI) technology can be applied to predict how vehicle consumer behavior changes to bring to judge whether vehicle consumer will like what kinds of vehicle styles next year. Then, vehicle manufacturers can gather overall vehicle consumer data to analyze and conclude the more accurate vehicle design direction for next year any new design vehicle manufacturing products.

Thus, (AI) machine learning can help vehicle manufacturers to solve how to design any new vehicle products challenge. A vehicle is both one of the most important and carefully considered purchases the majority of people will ever make in their lifetime. It is also a purchase that tends to be fundamentally tied to a person's identify and view of themselves. As the same time, vehicle consumers changing lifestyles result in changing vehicle needs, e.g. the young sport car enthusiast matures into the family driver.

Automotive dealers need to remember that vehicle customers and prospects are individual human beings with risk, complex and ever-changing lives factors, these factors will influence every vehicle consumer why who feels has vehicle purchase need, and how who choose to buy the first vehicle if who decided to buy the first vehicle.

The (AI) technological customer behavioral prediction tool seems to be the best vehicle salespeople in the world are those that know every one of their vehicle customers. Their likes and dislikes which style of vehicle design, preferences and changing tastes to vehicle choices. The capacity of the human brain, however, limits us from achieving this type of vehicle sales and frequent turnover at vehicle dealerships often results in the further loss of vehicle salespeople along with their vehicle customer relationships and knowledge. In this competitive vehicle environment, machine learning enables platforms to assist the vehicle sales team by tracking the vehicle consumer behaviors of each vehicle customer, learning and memorizing their preferences and predicting their future vehicle purchase needs.

Finally, I recommend that for a vehicle dealerships marketing platform to make their customer engagement efficient and fully-functional, I should be able to: applying (AI) tools to track every vehicle customer behavior across the web, connecting to a society of data sources, CRM, DMS, third-party, web vehicle brands, social email, click etc., aggregating and accurately cross-reference data from a variety of sources, leveraging this data to drive insights on a mass scale, as well as on an individualized basis, driving actions and automatically direct customer engagement via multiple channels based on where each customer is in their individual lifecycle.

How can (AI) provide businesses with better-informed decisions

I shall explain how (AI) technology can provide businesses with better-informed decisions to drive top-line growth, deliver meaningful experience for customers and smooth their path along the consumer journey. The widely understood definition of (AI) involves the ability of machines or computers to learn human thinking, reasoning and decision-making abilities.

A Narrative science study in 2015 year identified that (AI) was being used primarily in voice recognition, machine learning virtual assistants and decision support. This study also highlighted the many branches of (AI) and that techniques and their definition are used interchangeably. It is possible that (AI) can be used to gather big data , then to analyze to help businesses to predict consumer behaviors. For example, one of the most common techniques is machine learning, where algorithms are used to perform tasks

● How (AI) influences organizational change

Consequently creative and social intelligence will be in even greater demand as (AI) makes in management and the workforce. This development will represent a long term trend in labor markets , one characterized by intensifying demand and reward for social skills with a growing desire for creative capabilities, managers will seek to fashion of ideas and hypotheses from inside and outside of the enterprise to shape solutions to their most pressing business problems. Thus, (AI) will influence overall organizational team members who have chance to participate any decision to make more accurate business judgment.

Many managers mistakenly view judgment work as only an individual discipline, failing to appreciate that it can also involve decide interpersonal and organizational practices. In more complex settings, judgment is typically a collective outcome of individuals' and teams' diverse perspectives, insights and experiences. And often , the resulting choices are better informed than decisions that an individual would have arrived at on his or her own.

Thus, when any organizations apply (AI) technology to assist managers to gather data and ideas to make any judgment. In these cases, organizations can create the conditions for effective collective judgment by establishing structures , such as " shadow advisory boards" that prompt managers and employees to source and synthesize multiple perspectives. Thus, a traditional organization (firm) might freshen its thinking is t put together a shadow advisory board, comprised of young, digital people who can apply (AI) machine assistance to make judgment work more accurate whether related to people development, problem-solving or strategizing and innovating for considerable degrees of creative and social intelligence.

Thus, on the one hand, (AI) technology machine augmentation and automation can give these advantages to human (organization managers) , e.g. developing people and community, solving problems and collaborating, coordinating and controlling work, shaping strategy and leading innovation. Besides, on the other hand, the next generation managers need have these individual attitude to treat intelligent machines to be as colleagues.

When, judgment is a human skill, intelligent machines can accelerate human learning that supports it, assisting in data -driven simulations, scenarios and search and discovery activities. Focuses on judgment work, some decisions require insight beyond what data can tell them. This is the sweet sport for human judgment, the application of experience and expertise to critical business decisions and practices. Thus, managers will also need to find ways to learn how to use digital (AI) technologies to tap into the knowledge and judgment of partners, customer external stakeholders and role models in other industries after the (AI) machine had been implemented to the organization.

Future works change:Automation, employment and productivity

2.1 How (AI) influences employment

Human future " micro to macro" industry trends will be affected business strategy and public policy by (AI) technology. In the future (AI) technology will influence those six themes: productivity and growth, natural resources, labor markets, the evolution of global financial markets, the economic impact of technology and innovation and urbanization. However, (AI) technology will bring economic benefits of tackling gender inequality, a new global competition, Chinese innovation and digital globalization.

Nowadays, advances in robotics artificial intelligence, and machine learning are in a new age of automation, as machines match or outperform human performance in a development to any countries. For example, automation of activities can enable businesses to improve performance by reducing errors and improving quality and speed, and in some cases achieving outcomes that go beyond human capabilities. For example, some research indicated automation could raise productivity growth globally by 0.8 to 1.4 % annually; more than 2,000 work activities across 800 occupations. When less than 5% of all occupations can be automated using demonstrated technologies about 60% of all occupations have at least 30% of constituent activities that could be automated. Many occupations will change that will be automated away: Activities most susceptible to automation involve physical activities, in highly structured and predictable environments, as well as the collection and processing of data. They are most prevalent in manufacturing , accommodation and food service and retail trade and include some middle-skill jobs. For example, such as natural language processing is a key factor. Beyond technical feasibility, the cost of technology competition with labor including skills and supply and demand dynamics, performance benefits including and beyond labor cost savings, and social and regulatory acceptance will be affected by (AI) automation technology. Thus, (AI) automation will impact to influence global employment in those aspects as below:

Firstly, assuming that people are displaced by automation will find other employment. The anticipated shift in the activities in the labor force is of a similar order as the long-term shift away from agriculture and decreases in manufacturing share of employment. Both of manufacturing and agriculture industries which would be accompanied by the creation of new types of work not foreseen at the time.

Secondly, for business, the performance benefits of automation are relatively clear. Thus, the businessmen have opportunities for their micro economies to benefits from the productivity growth potential and macro economies to benefit to encourage continued progress and innovation , investment and market incentives. At the same time, employers must innovate policies to help workers and institutions adapt to the impact on employment.

This will likely include rethinking education and training, income support and safety nets , as well as support for those dislocated, when employees need to leave themselves homes to move to other cities to learn new (AI) automation works. Thus, individuals in the workplace will need to engage move comprehensively with machines as part of their everyday activities, and acquire new skills that will be in demand in the new automation age. Consequently , the scale of shifts in the labor force over many decades that automation technologies can be a similar order to the long -term technology -enables shifts in the developed countries' workforces away from agriculture in the 21 th century. Those shifts did not result in long-term mass unemployment because they were accompanied by the creation of new types of work not foreseen at the time. However, human will still be needed in the workforce when the total productivity gains are caused by

(AI) technology.

2.2 What occupations will be influenced by (AI) technology.

In the future, scientists predict that these occupations will be influenced by (AI) technology mostly. They include : retail salespeople, food and beverage service workers, language or translation teachers, health practitioners. Since these work activities have a more relevant occupations are made up of a range of activities with different potential for (AI) automation . For example, a retail salesperson will spend more time interacting with customers, stocking shelves , or ringing up sales. Each of these activities is distinct and requires different capabilities to perform successfully.

Thus, these job activities have similar simple control characteristics. Simple activities include greet customers, answer questions about products and services, clean and maintain work areas, demonstrate product feature process sales and transactions. All these activities can have similar simple activities in order to (AI) machines can be learn how to do these activities from (AI) technology . For example, the capability perception includes sensory perception, cognitive capabilities, such as retrieving automation, recognizing known patterns(supervised learning), logical reasoning problem solving.

Thus, (AI) machine is such human, which has feeling and emotion, such as social and emotional sensing, judgement reasoning methods, natural language understanding and physical capabilities, such as mobility , navigation, gross motor skill, fine motor skills. It seems that the future, (AI) human invents machines which will have these human characteristics to do human similar behavioral job duties more easily and efficiently. It implies these above human occupations will be replaced by (AI) human invention machines in the future. Due to (AI) creation, it is possible to cause unemployment number of these above workers will increase because (AI) machines can do their similar job behavioral activities.

Consequently, employers won't need to employ many of these skillful labor. Otherwise, they can buy less number (AI) machines to attempt to do whose job activities more easily and efficiently. So, it seems (AI) machines will have more high work performance to replace these occupation workers' work performance. Finally, these occupation worker unemployment number will only increase when the (AI) machines had been invented to achieve to do their work behavioral activities absolutely success in the future.

2.3 Whether (A) technology machine labor
will replace human worker more or assist
human worker more

There is no single agreed definition of a robot how outcome of a task that is completed without human intervention. When some definitions require the task to be completed by a physical machine moves and respond to its environment, other definitions use the term robot in connection with tasks completed by software , without physical embodiment.

However, to answer the question : Whether (AI) technology machine labor will replace human worker more or assist human worker more. I shall indicate some examples to let readers to judge whether (AI) technology can create new jobs or reduce old jobs.

Firstly, I shall explain what (AI) function is. (AI) is a service robot that performs useful tasks for humans or equipment excluding industrial automation application . Thus, the classification of a robot into industrial robot or service robot is done according to its intended application. It is also a personal service robot or a service robot for personal used for a non commercial task, usually by lay persons . Examples are domestic servant robot, and pet exercising robot. It is also a professional service robot or a service robot for professional used for a commercial task, usually operated by a properly trained operator. Examples, are cleaning robot for public places, delivery robot in offices or hospitals, fire-fighting robot, rehabilitation robot and surgery robot in hospitals. Thus, these functions will be future (AI) application to our daily life necessaries or business necessaries.

However, some authors agree (AI) will bring negative outcomes of automation, due to raise competiveness, reduce human job nature. Otherwise, other authors argue (AI) will bring positive outcomes of automation, due to raise productivities, job creation, assist humans work.

On the positive outcome hand, robots can increase productivity . This is particularly important for small-to medium sized businesses both are in developed and developing countries economies. It also enables large companies to increase their competitiveness through faster product development and delivery. Increased use of robot is also enabling companies in high cost countries to re shore, or bring back to their domestic base parts of the supply chain that will have previously outsourced to sources of cheaper labor. Currently , the greater threat to employment is not a automation, but an inability to remain competitive. Automation has led overall to an increase in labor demand and positive impact on wages. The reason is that the middle-income/middle-skilled jobs have reduced as a proportion of overall contribution to employment and earnings leading to fears of increasing income inequality, the skills range within the middle income bracket is large. Thus, robots are driving an increase in demand for workers at the higher -skilled and with a positive impact on wages. This issue is how to enable middle-income earners in the lower-income range to unskilled or retain. Finally, the (AI) positive impact supporter who argue the future will be robots and humans can work together.

However, on the negative outcome hand, robots can substitute labor activities, but don't replace jobs. They believe that less than 10% of jobs are fully automatable. Increasingly , robots are used to complement and augment labor activities, the net impact on jobs and the quality of work is positive. Automation can provide the opportunity for humans to focus on higher-skilled, higher-quality and higher-paid tasks. Robots can improve productivity when they are applied to tasks that which perform more efficiently and to a higher and more consistent level of quality than humans. For example, increased productivity is enabling some firms, such as Whirlpool, Caterpillar and Ford Motors company in the US restructure their supply chains, bringing back parts of the manufacturing process to the country of origin. Thus, productivity gains due to robotics and automation are important not just at the company level, but also for build industry and nation competitiveness.

I suppose that productivity can be raised. What are the impacts of robots on employment? Firstly, the main focus of development has been on personal entertainment, which does not drive worker productivity (manufacturing production). When the internet (information and communication technology (ICT)) innovation. This is borne and by findings that manufacturing productivity, which has been driven by innovations in automation rather than consumer technologies, has government strongly than productivity in the services sectors of the economy in most nature economies. It seems (AI) automation will create many jobs in internet communication entertainment game industry. For example, many young people like to use internet to play any electronic games from computer or mobile at home or outside home conveniently. Thus, (AI) automation will increase demand to be invented to any new entertainment game from internet channel. It will need to employ many (AI) entertainment game inventors to create many automation entertainment games. Thus, (AI) automation in internet entertainment game industry will need human (AI) entertainment game inventors to invent the knowledge-based capital of (AI) automation entertainment games. The (AI) entertainment game inventors will need own research and development skills, form specific skills, organizational know-how skills, databased knowledge, design and various forms of intellectual property to do these (AI) automation entertainment game invention occupations in the future.

International Federation Of Robotics(2016) indicated that China will be as a major robotics manufacturer and user of robots, benefiting from jobs created by robot manufacturing and productivity gains from robot use. Chins had sold of robots to any one single market every year since 2017 year. The Chinese government has included a focus on robotics in its 10 year strategy. In order to achieve its target of a robot density of 150 units per 10, 000 workers by 2020 year. Thus, Chinese companies will have to install around 650,000 new industrial robots between 2016 to 2020 year, 2.5 times more than installed globally in 2015 year.

Hence, China (AI) manufacturing industry will need to employ many workers . It implies (AI) manufacturing industry will create many new occupations in China. Also, ministry of economy, trade and industry (2015) also showed that Japan currently has the largest stock of industrial robots in operations, primarily in the automation industry. Driven by a rapidly aging population and low productivity rates, the Japanese government has sights on a 20-fold increase in the use of robots in the non-manufacturing sector and a three-fold growth rate of labor productivity in the service sector both by 2020 year. Thus, it also implies Japan will need many robots to be provide to service industry. Due to robots will provide to serve any businessmen's clients. Thus, it is possible that the service workers won't be dismissed as well as it is depended on the serving job nature to decide whether Japan's service workers can still serve to their employer when the service (AI) robots are applied to whose employers.

Consequently, it seems that (AI) can create employment, Ministry of economy, trade and industry (2015) showed that such as China will develop the major (AI) automation manufacturing industry. The (AI) employers will need to employ many workers to manufacture any these different kinds of (AI) robots to satisfy China or overseas individual or business buyers needs. But, (AI) can also cause unemployment to the low skillful service workers. Such as if Japan some service businesses choose to buy any (AI) service robots to replace their service staffs to serve their clients. It is possible that the service staffs will be dismissed, due to (AI) robots can do such as their same service job duties to achieve better service performance. Thus, today, it is increasingly common for people to use robots in various situations at home and in retail stores, hotels and hospitals these service industries. Robots are classified into server types based on their functionality (service and utility robots or those designed to communicate with humans) and appearance (humanoid robots or mechanical robots). The type of robot, to which each country allocated particular importance in the advance of robotics, reflects the sense of values and preferences of its population. Thus, if the country has high population needs to use robots, then they will influence either more new jobs creation or more old job loss in the country's (AI) manufacturing or (AI) service industries both. For example, Japan respondents often associate the term " robot " with humanoid robots that can communicate with human and they have a high level of familiarity with robot. The US has the highest level of robot utilization at home and in retail stores with its people being the most enthusiastic about the future use of robots. Germany shows a strong tendency to consider robots for industrial purposes and its people feel strong effort to the presence of robots in their households.

In conclusion, to judge whether how (AI) will influence the country's employment to be better or worse. It will depend on the country home buyers (users) or business buyers (users) how to use (AI) for their daily needs. If the country , such as US retail stores need to use (AI) , it will have possible to reduce some or many retail service workers. Even, if the country , such as Japan has many home users need to use (AI) , it will not influence the employment market. Otherwise, it will raise (AI) salespeople numbers. Even, if the country, such as Germany and China will have many (AI) manufacturers, then it will create many (AI) manufacturing occupations for these (AI) manufactory workers.

Consequently, (AI) robots manufacturing and service needs will have positive or negative impact to any country's employment. It will depend on the (AI) service provision and service workers' job nature as well as the manufacturing workers of (AI) knowledge level to decide their employment chance in their country's employment market.

AI brings what jobs change
   What does artificial intelligence(AI) mean?
● What (AI) function is?

Some scientists explain that artificial intelligence means which is an expert system, computer software that embodies a portion of the specialized knowledge of a human portion in a specific, narrow domain, owns decision making ability of human expert. The (AI)technology is based on the premise that what makes a person an expert is years of experience that enables who recognizes certain patterns in a problem as being similar to pattern. For example, in the future artificial intelligence system can be applied to control air traffic, design to computer configuration, medical diagnosis, instruction/training, speech/interpretation, monitoring to (nuclear plant), planning to mission, factory scheduling, prediction weather, repairing telephone, automatic driving etc. different industries.

Artificial intelligence characteristics include: creative, adaptive , common sense, fact processing, quick replication, broad focus permanent and consistent skill. Otherwise, traditional computer expert system disadvantage includes perishable, unpredictable, slow reproduction, expensive, slow reproduction, slow processing lacks inspiration, needs instruction, narrow focus only machine knowledge. So, artificial intelligence is a branch of computer science devoted to creating computer to influence software and hardware to attempt to create human intelligence or human intelligent behavior. It is learning from experience, responds flexibility in situation that are, new or not anticipated.

Thus, (AI) can be learnt programmed knowledge to solve problems, using reasoning in solving problem, understanding and inferring facts and rules, recognizing the relative importance of different elements in a situation. In summary, artificial intelligence is concerned with two basic ideas mainly: The first idea, it involves studying the thought processes of humans to understand what intelligence is; the second idea, it deals with representing thought processes using companies to create artificially intelligent entities for testing the theories of intelligence.

   ● Can (AI) impact human job nature?

Human need concern this question: Will artificial intelligence (AI) reduce some human jobs in order to instead of replacing machines to do? Due to artificial intelligence is the ability of machines to do thing, that people would require intelligence. For example, artificial intelligence machine man driving(self-driver), it (AI) machine man driving research is an attempt to discover and describe aspects of human intelligence that can be simulated by driving machine functions. Alternatively, (AI) mathematical research may be another viewed as an attempt to develop a mathematical theory function to describe the abilities and actions of things (natural or man-made) exhibiting intelligent behavior and server as a design of intelligent calculation machine function.

Why do humans need artificial intelligence machines to instead of traditional human service job? For example, can artificial intelligence machine man (self-driving) driver drive to replace human driver? I shall compare the differences between humans and computers : The characteristics of humans are good at recognizing various things, either seen before or not, recognizing the relationship patterns between things. Human thinking is common sense reasoning, combining all types of sensory input, acting appropriately in novel situations, learning new things and changing behavior patterns, making decisions , even when given incomplete information, working with noisy, incomplete information gathering behaviors . However, characteristics of computers are good at: The tasks humans do naturally are extremely difficult for a computer program as intelligent, which must be able to do the same kind of tack as humans do naturally.

Hence, (AI) is an combination of many different success and technologies: Linguistics - computational and socio, philosophy-logic, philosophy of mind and of language, electronical engineering -image and speech processing, pattern recognition, robotics, machine learning, neural networks, optimization scheduling, management information system and decision making. So, it is possible that (AI) can impact human job nature to instead of human working behavior in the future.

How can (AI) influence labor market?
● How can human society job nature
to be changed to artificial intelligent society?

From the first intelligent perspective reason view point, artificial intelligence is making machines " intelligent" acting as humans expect people to act. Artificial intelligence has ability to distinguish computer responses from human responses, it owns knowledge to solve expert problem. From another research perspective reason view point, artificial intelligence is the study of how to make computers do things which, at the moment, people do better (Rich & Knight, 1991, p.3).

(AI) researchers are native in a variety of domains, e.g. formal tasks (mathematics, games), tasks (perception, robotics, natural language, common sense reasoning), expert tasks (financial analysis, medical diagnostics, engineering, scientific analysis and other areas).

From the second business perspective reason view point, (AI) is a set of many powerful tools, and methodologies for using those tools to solve business problems. From a programming perspective reason view point, (AI) includes the study of symbolic programming problem solving and search .

From the third human technological perspective reason view point, today's computer can do many well-defined tasks, for example, arithmetic operations, are much faster and more accurate than human beings. However, the computers' interaction with their environment is not very sophisticated yet. How can human test whether a computer has reached the general intelligence level of a human being? Can a computer convince a human interrogator that it is a human? But before thinking of such advanced kinds of machines, human will start developing our own extremely simple " intelligent" machines.

So, it is possible that human society job nature will to be changed to artificial intelligent society when (AI) technology is developed to the mature stage in the future.

● Why does human need artificial intelligence machines?

One of major division in (AI) is between humans who think (AI) is the only serious way of finding out how we (human) work and human who want companics to do very smart things, independently of how we (human) work. This is the important distinction between cognitive scientists vs engineers. One of another major division in (AI) is between symbolic (AI), which represents information through symbols and their relationships. Specific Algorithms are used to process these symbols to solve problems or deduce new knowledge and connectionist. So (AI) , which represents information in network. Biological processes underlying learning, task performance and problem solving are imitated from human mind behaviors. Thus, it is possible that artificial intelligence machines can do the better judgicious behavior to compare human.

● How does artificial intelligence influence future working changing in automation employment and productivity aspects?

In the automation changing influence aspect, as companies increasingly use robots on production lines or algorithms to optimize their logistics manage inventory, any carry out other core business functions. Technological advances are creating a new automation age in which ever-smarter and more flexible machines will be deployed on an ever larger scale in the marketplace. However, researching artificial intelligence with how influences human working nature. We need to answer these questions: How will automation transform the workplace? What will the implications for employment? And what is likely to be its impact both on productivity in the global economy and on employment?

Advances in robotics, artificial intelligence, and machine learning are growing in a new age of automation as machines match or outperform human performance in a range of work activities, including ones requiring cognitive capabilities. What factors are determined the changing in workplace adoption by artificial intelligence innovation? What advantages are automation? Automation of activities can be enabled businesses to improve performance by reducing errors and improving quality and speed, and achieving outcomes that go beyond human capabilities.

Some scientists indicated based on their scenario modeling. They estimated automation could raise producing growth globally by 0.8 to 1.4 percent annually. Almost, the activities people are paid almost $16 trillion in wages to do in global economy have the potential to be automated by adopting currently demonstrated technology. According to their analysis of more than 2,000 work activities across 800 occupations. When less than 5% of all occupations have of least 30% of activities that could be automated. They also indicated that technical economic and social factors will determine automation. Continued technical progress, for example, in areas such as natural language processing is a key factor beyond technical feasibility , the cost of technology, competition with labor including skills, and supply and demand dynamics, performance benefits including and beyond labor cost savings and social and regulatory acceptance will affect (alter) the scope of automation.

Other some scientists also indicate U.S. country for example, the anticipate shift in the activities in labor force of a similar order of magnitude as the long term sight away from agriculture and decreases in manufacturing. Share of employment in the United States both which were achieved. So, those factors can influence why artificial intelligence technology needs. So, it is possible that future agriculture and manufacturing both industries will apply (AI) technology manufacturer-kind of job nature to raise productivity instead of farmers, fruit picking workers, farming transportation labours as well as factory manufacturing workers and supervisors etc. human-kind of job nature.

● Is artificial intelligence possible to replace labor ?

Not just intelligence, but also debating, if machines are capable of having a conscious minds. Artificial intelligence has those characteristics as below:

On functionalism aspect, artificial intelligence inputs mental states, sensory inputs, (beliefs, desires being in pain feeling) and behavioral outputs. Since mental states are identified by a functional role, which are thoughts to be manifested in various systems. Even, perhaps computers which are physical devices with electronic substrate that inform computations on inputs to give outputs similar to brains which are artificial intelligence composed of part any intrinsic relationship to each other. Thus, artificial intelligence activities is not the whole itself, but into parts or on external influence on the parts.

On dualism aspect, artificial intelligence is a set of views about the relationship between mind are matter. On materialism aspect, it builds the only thing that exists is matter, including consciousness.

On biological naturalism aspect, it is similar a human brain than feels pains makes mental situation. So, artificial intelligence is similar biologist which might to be excited to human labor work. Hence, it seems artificial intelligence can change (alter) or replace human labor work of nature in possible in the future.

● Can (AI) technology replace human labour nature of work?

On technological innovation reason view point, the history development of artificial intelligence studying the intelligence is one of most ancient scientific discipline. The history development of artificial intelligence what aims to achieve human use to sense, learn remember and think, logic probability, decision making and calculation develop from mathematics, instead of replacement human labor functions.

Artificial intelligence history development aim is the scientific analysis of skills in connection and practice with the appearance of computers from 1950 year beginning. The artificial intelligence (AI) can deal with the ultimate challenges. How can (either biological or electronic) mind sense, understand and manipulate a world that is much simple and more complex than itself? And what if would human like to construct something with such capabilities?

The general-purpose software of the early period of (AI) were only able to solve simple tasks effectively and failed when which should be used in a wider range or an more difficult tasks. One of the sources of difficulty was that early software had very few or mix knowledge about the problems which handled, and activities successes by simply syntactic manipulation. Moreover, the other difficulty was that many problems that were tried to solve by the (AI) were untreatable.

The early (AI) software whether trying step sequences based on the basic facts about the problem that should be solved, experimented with different combinations till which found a solution. From the end the 1960 year, developing the so-called expert systems were emphasized. These systems had (sue-based) knowledge base about the field which handled. Till to the beginning of the 1970 year, (Prolog) the logical programming language was born, which was built in the computation realization of a version of the resolution calculus. (Prolog) is a remarkably prevalent tool in developing expert systems (on medical, judiciary and other scopes), but natural language parsers were implemented in this language. Then, in 1981 s, the Japanese announced the fifth generation computer system project, a 10 years plan to build an intelligent computer system that use the (Prolog) language as a machine code. Nowadays, (AI) can be applied any industries, such as car manufacturing industry can use (AI) technological machine-men manufacture car, instead of replacing human labors in factory. Even, in the future, using (AI) machine-men drivers can drive any private cars or public transportation tools, instead of replacing human drivers, e.g. bus, train, tram, ferry etc. Also in the future, machine-men can replace housewives to serve families to do housekeeping clean job , e.g. cleaning toilets, bathrooms, kitchens, even cooking functions at home. So (AI) machine-man can reduce housewives works at home. Moreover, (AI) machine man can take care old people , when who are living at homes or elder care centers.

So, it seems artificial intelligence (AI) will be possible developed to manufacture a new generation machine-man to assist (serve) families to do any simply cleaning or cooking jobs at homes. Moreover, the overall demand of (AI) general social needs will also rise, such as security, driving transportation tools, restaurant cleaning, elder centers care service etc. So, it seems that individual or families or social needs of (AI) will be increase in the future. Thus, it will influence macro economy growth (GDP) if there are large house family consumer group and hotel or bus or taxis or ferry etc. different business consumer group demand any artificial intelligence machine numbers increasing. Then, the artificial intelligence products and material manufacturers must need to buy many artificaial intelligence materials to produce any kinds of artificial intelligence machines to prepare to satisfy consumer individual needs. Consequently, macro economy will grow to the owned artificial intelligence development countries, e.g. US, China, UK.

● Why can artificial intelligence satisfy human needs?

First, On machine-man satisfactory demand aspect view point, it makes computers that think, it is the automation of activities. We associate with human thinking: like decision making, learning. It is the act of creating machine that perform function that require intelligence when performed by people. It is the study of mental faculties through the use of computational models. It is the study of computations that make it possible to perceive, reason and act. It is a branch of computer science that is concerned with the automation of intelligent behavior. It is anything in computing service that human don't yet know how to do property.

Second, on thought aspect artificial intelligence means systems thank think like humans, systems that think rationally.

Third, on behavioral aspect, artificial intelligence systems that act like human and that systems act rationally. However, the basic objective of (AI) is to represent human's thought processes in computation . These machines are supposed to exhibit behavior that. It is performed by a human being, would be considered intelligent. However, some authors feel (AI) has disadvantages, such as it is not creative, it is excited in the use of sensory devices, it can't make use of a very wide context of experiences and it does not use common sense.

For speech recognition and understanding function needs example, (AI) can be applied in speech recognition and understanding function, which (AI) speech or voice recognition is a data input method. For example, the computer recognizes and understands one (or a few) word commands. Speech understanding on the other hand is the computer's ability to understanding a spoken language. That is , the computer understands the meaning of sentences, an paragraphs through (AI).

So, (AI) can be attempted to learn human language how to speak. It is similar to translate human language skill, instead of actual human speaking skill. Also, (AI) can assist handicap learning or language student how to listen different languages by machine-man sounds from computers more accurately. So, it seems that it (AI) can replace human language teachers speaking function and can change teaching language nature of job in language speaking and listening education industry.

● Is artificial intelligence one good choice for human future technological benefit?

Nowadays, new technology development is popular. However, artificial intelligence is one kind of new technology choice among different technologies innovation. So it brings this question: Is artificial intelligence technology value to invest? To answer this question. I shall indicate some other new technology developments to compare (AI) technology development to judge which has urgent needs to achieve human expectation nowadays.

For example, why is green peace interested in new technologies? New technologies features prominently in our ongoing campaigns against genetic modified crops and number power. However, which are also an integral part of our solutions to environmental challenges, including renewable energy technologies, such as solar, wind and wave (water) power energy as well as waste treatment technologies, such as mechanical, biological treatment.

It seems humans need concern how to apply (AI) technology to solve environment pollution challenges in our future. So, environment protective, agriculture, natural energy technology will be popular demand to attempt to apply (AI) technology to solve their challenges or apply (AI) to assist to develop their industry.

What is influence to (AI) job nature change to change developed and countries economy

● How can artificial intelligence technology influence economy?

Advances in artificial intelligence (AI) technology and related fields have opened up new markets and new opportunities progress in critical areas, such as health, education, energy, economic development, social welfare and the environment pollution.

(AI) automation will continue to create wealth and expand the global economy development in the future. However, when many will benefits that growth won't be costless and will be accompanied by changes in the skills, that workers need to increase productivity in the economy and structural changes in the economy. So, in the skills that workers need to succeed in the economy and structural changes.

I shall indicate why aggressive policy action will be needed to help Americans who are disadvantaged by these changes , due to (AI) technology is caused. For automation industry change example, artificial intelligence (AI) capabilities will enable automation of some tasks that have long required human labor. These artificial intelligence technology introduction can increase new opportunities for individuals. The economy and society, but (AI) has also the potential to disrupt be current livelihoods of many Americans. However, (AI) leads to unemployment and increase in inequality over the long run depends not only on the (AI) technology itself, but also on the institutions and policies that are changed.

Thus, it is possible that (AI) technology will raise some countries unemployment number if the employer apply (AI) technology workers to work instead of human labor in their factories, but it can also raise productivities for these employers.

● Can (AI) influence global economy growth?

Technological progress is main driver of growth of GDP per capita, allowing output to increase faster than labor and capital . However, technology can increase productivity, but also decrease the number of labor hours needed to create a unit of output. So (AI) causes unequal to labor wage decreases, even reduces the number of labor to manufacture, e.g. artificial intelligence technology of automation car manufacturing industry; clothing manufacturing industry; plane manufacturing etc. high technology of artificial intelligence manufacturing method. But (AI) should be potential environment benefit, although it raises unemployment ratio. Moreover, it can rise production , due to many skilled craft were replaced by the combination of machines and lower-skilled labor. The result of (AI) technology introduction , it causes output per hour risen when inequality declined, driving up average living standards, but the labor of some high-skill workers was no longer as valuable in the market. Otherwise, if (AI) technology is continue developed to be success. Some routine intensive occupations will be loss, which focused on predictable, e.g. easily programmable tasks, such as switchboard operators, filing clerks, travel agents, and assembly line workers would be particularly replaced by new (AI) technology. However, at the same time, (AI) technology development will bring these benefits: improvement in education (training (AI) technology scientists) , due to (AI) manufacturing technology needs are raising to businesses and institutional changes, such as the reduction in unionization and raising in the minimum wage to the (AI) manufacturing technology skilled labor in factories.

Because (AI) technology is not a single technology, but rather a collection of technologies that are applied to specific tasks, the effects of (AI) will be felt unevenly though the economy. It will bring some tasks will be most easily automated than others , and some jobs will be affected more than others, both negatively and positively. Finally, new jobs are likely to be directly created in areas , such as the development and supervision of (AI) as well as indirectly created in a range areas though out the economy as higher incomes lead to expanded demand.

However, if (AI) technology could dominate global labor markets. If labor productivity increases, do not influence into wage increases, then the large economic gains brought about by (AI) technology could be increased wealth inequality, due to employers can reduce production cost, but workers (labors) wages will not be increased, even will be decreased. Hence, it seems the (AI) technology will bring disadvantages to labor market to cause unemployment or reduce wages in possible, although it can reduce employer individual salary (wage) expenditure and it can raise productivity.

● How can artificial intelligence impact global economy growth?
Artificial intelligence (AI) technology is a branch of computer science that aims to create intelligent machines that work and react like humans. So, (AI) is a technology that appears to impact (influence) human preference by learning, understanding complex contents, enhancing humans in executing both routine and non-routine tasks. In the future, (AI) technology that can be virtual personal assistant, as well as it may exist, such as robots with human-like processing capabilities.

How can (AI) technology impact global economy growth over the next 10 years? During this time period, (AI) technology is predicted to have wide-ranging applications including: Machine learning that automates analytical model building by using algorithms that allow machines to operate without human assistance.

In global education aspect, potential applications include predicting cause-and-effect relationships from biological data, identifying new drugs, self-driving cars, and protecting against fraud, improved natural language processing that allows computers to continue to better analysis, understand and generate language to interface with humans using natural human languages. For example, transcribing notes dictated by physicians, automatically drafting articles and translating text and speech. So (AI) technology can be applied to education aspect to improve humans' knowledge level.

In visual art aspect, (AI) machine vision that allows computers to identify objects, scenes and activities in images. Current applications of (AI) machine vision include providing objective descriptions for the blind seeing(visual) needs.

We except the economic effects of (AI) technology to include both direct GDP growth from sectors that develop or manufacture. (AI) technology and indirect GDP growth through increased productivity in existing sectors that employ some form of (AI). If (AI) technology is an increasingly critical component of more products, it will become an integral part of many people's lives. Thus, (AI)'s ability to influence economic activity, rather than the economic or development status of the region. (AI) has the potential to impact income classes and to bring significant gains to both developed and developing countries. For example, (AI) has the potential to optimize good production around the world by analyzing agricultural regions and identifying what is necessary to improve crop yields.

In estimating the future economic effects by (AI) technology innovation, it is important to note that it is challenging to accurately predict which applications of (AI) will ultimately be commercially successful. In micro level economic influence, we need to apply methodologies to estimate the economic effects of investment in firms developing (AI) technology since investment levels in a technology are a telling sign of the future potential of that (AI) technology.

● How can (AI) influence GDP of high income countries in the next ten years?

How (AI)'s development may affect the global economy over the next ten years. In fact, (AI) technology has the potential to affect business across the global in a wide range of industries in ways only a number of technologies have done in the parts. For example, (AI) technology's expected to be a useful tool for enhancing human capabilities and in some instances replacing functions, such as driving a car, adoption of broadband internet, mobile telephone, industrial robotic automation have served to enhance human capabilities.

However, significant public debate has focused on projections of (AI) technology's effect on the labor force. However, large companies prefer to invest in (AI) technological industry. For example, face book's (AI) research lab., google machine intelligence lab. and micro soft machine learning and artificial intelligence research division are all making advances in (AI) technology and investing in the industry's top talent. Additionally, between 2010 year and 2015 year, nearly \$5 billion in venture capital funding invested in firms across the global developing and employing (AI) technology (Facebook (AI) Research).

● How can artificial intelligence impact on workplace?

Modern information technologies and the labor economy growth of machines is powered by artificial intelligence have already strongly influenced the world of work in the 21 ST century. Computers, algorithms and software simplify every tasks and it is impossible to image how most of our life could be managed without them. How can be the information economy characterized by exponential growth replaces the most production industry based on economy of scales? What will the future world of work look like and how long will it take to get? Will the future world of work be a world where humans spend less time earning their livelihood? Alternatively, are mass unemployment, mass poverty and social distortions also possible scenario for the future, where robots, artificial intelligence systems play an increasingly central role? These questions concern how artificial intelligence further development . Can influence labor economy growth on workplace ? When the labor market has widespread impact on intelligence property, information technology, product liability, competition and labor and employment laws.

How (AI) technology impacts on labor workplace.

The future influence any organizations how labor economies use of (AI) can be analyzed, such as deep machine learning is based on a set of model high level data. Unlike human workers, the machines are connected the whole time in workplace. If one machine makes a mistake, all autonomous systems will keep this in mind and will avoid the same mistake the next time.

Over the long run intelligent machines will win against every human expert. Production robots have been replacing employees because of the (AI) technology. They work more precisely than humans and cost loss. Creative solutions like 3D printers and the self learning ability of these production robots will replace human workers, the automatic data recording and data processing, traditional back office activities are no longer in demand. Autonomous software will collect necessary information and will send it to the employee who needs it. Additionally, dematerialization leads to the phenomenon that traditional physical products are becoming software. For example, CD or DVDs are being replaced by streaming services. The replacement of traditional event ticket, e-travel ticket service products or hard cash will be the next step, due to the possibility of payment by smartphone. So, (AI) technology will impact human's daily life consumption behaviors in the future. For another example, transportation tools, such as boats and ferries and private vehicles will use sensors and navigating without human input. Taxi and truck drivers will become obsolete, the stock store applies to stock managers and postal carriers of the delivery is distributed by (AI) machine delivery method.

What is the relationship between (AI) and (CRM)?
● Can (AI) technology impact on customer relationship management (CRM) ?
Nowadays , (AI) is a technology almost as old as the computer industry itself, it is similar with the advent of personal assistants function to businesses and personal promotion channel, such as (Amazon's Alexa, Apple's Siri, Google's Assistant) image recognition (face book), personalized recommendations (Netflix , Amazon). Those innovations have been driven by a increase in processing power, lower cost hardware, and the exploding creation and availability of data. It seems, (AI) technology can impact global customer service management method.
How to forecast economic impact modeling to (AI) will affect global economy? Can human forecast business revenue growth and job creation (or destruction) based on (AI) applied to customer relationship management (CRM) activities? In addition to the economic impact on (AI) or (CRM) which can include an estimate of the economic impact attributable to sales forces customer base. What can economic benefits be brought to (CRM) from (AI) technology?
Artificial intelligence(AI) comprises a set of technologies that use natural language processing, machine learning, knowledge graphs, and other tools to answer questions, discover insights and provide recommendations. Computer systems can use (AI) hypothesize and formulate possible answers based on available evidence can be trained through the ingestion of vast amounts of content, and automatically adapt and learn from (AI) self mistakes and failures.
So, any business organizations (customer service departments) can provide efficient and effective customer relationship management of excellent customer service quality if which applied (AI) technology system. The different type of (AI) systems include: (AI) system platforms, machine learning (AI) based data preparation and enrichment tools, machine vision/

image recognition, voice speech recognition, text analysis and natural language processing, bots , e.g. face book website and virtual digital assistance solutions, social media pattern analysis , sentiment analysis, advanced numerical analysis (e.g. IOT streaming , machine logs), supporting technologies, knowledge base dialog management, Q&A processing etc. different (AI) technology system customer relationship management (CRM) tools.

(AI) (CRM) of activity can include these categories, such as: corporate marketing, marketing operation, field marketing, customer support, digital commerce, customer analytics, customer influenced product or service design, product or service pricing, finance information, presentation, customer billing, inventory , logistics and fulfilment support, partner management etc. different CRM tools.

(AI) technology of CRM has been carrying on plan different stages to achieve CRM personal assistant tool for businesses. The stages are such as, in the beginning stage of (AI) projects in place, implement now, pilot phase next year in the final stage of (AI) customer relationship management tools are foreseeable future. So, this CRM technology has been improved to plan in different stages every year to prepare to achieve full capacity of CRM service quality for businesses to use in the future.

Hence, how to develop an estimate prediction of the economic impact (AI) technologies could have CRM activities, which depends on gathering macroeconomic information on business revenue and the basic marketing of business revenue and the basic markup of business expenses by major functions (customer support, marketing and sales , production etc.)

An economic impact model that can gather data together and forecast the results how (AI) artificial intelligence technology brings (CRM) customer relationship management benefits to businesses, e.g. surveys investigation includes IT spending by sample countries, GDP and population estimates and forecasts, revenue per employee and ratios of IT spend to GDP. Surveys (questionnaire questions) of forecast results are influenced by (AI) impact can include: results are projected from surveys and rely on estimates are made by respondents on the expected financial improvements in categories of (AI) –assisted customer relationship management activities. The forecast assumes that these estimates are correct; financial estimates are based on estimates of "first year" improvement from full (AI) implementation; forecasts are from planning to implement any artificial intelligence of customer relationship management (CRM) projects, the improvement forecast is of categories of activity , e.g. corporate marketing , digital commerce, and customer analytics. They are not estimates of ROI for the (AI) software. They rely on conservative estimates to which each of these entities might affect company revenue, expenses or productivity. They also rely on estimates of the penetration of software in customer relationship management activities . Net new jobs created are based on the ratio of new revenue to jobs required to support that revenue . They can assume that 50% of the net new revenue will support increases in labor and the rest will go for capital and other operating expenses that may replace jobs lost to automation.

In the future, some of the ways in micro economic benefits to any organizations. (AI) technology is expected to impact CRM activities include: Spending up sales cycles, improving lead generation and qualification solving customer support problems faster (raising service quality), helping companies improve brand campaigns and recognition, lowering costs of support calls when increasing resolution rates, lowering the cost of recruiting employees and partners, increasing revenue from optimized product marketing, optimizing price, distribution logistics and preventing loss through fraud detection. So, micro economic benefits view point, it seems that (AI) CRM technology can raise any companies economic benefits for care term.

Artificial intelligence enables machines or the in-build software to behave like human beings which allows these decisions and act. The advent of (AI) is leading , talking, making decisions and act. The advent of (AI) is leading to new technologies advances and transforming the economic and employment opportunities for humans in a positive way. (AI) related technologies can facilitate our live. For example, industrial robotics, robotic medical assistants, smart games, financial forecasting software, big data analysis, algorithms in health and bioinformatics, pilotless cargo places, drone ambulances and general purpose and workplace robots and others. (Disruptors technologies: Advances that will transform life, business and the global economy).

Artificial intelligence also known as computational intelligence is defined as " the human –like intelligence exhibited by machines or software. It is theorized that intelligence of humans can be described and intelligence machines or software can simulate it. These machines software can be reasonable , learn, perceive and process information, like human mind and thus facilitate human life. They can think and act for us. So, artificial intelligence is an interdisciplinary field of study including computer science, neuroscience, psychology, linguistics and philosophy.

However, (AI) research and developments have economically impacted many industries, such as robotics, telecommunications, computer applications , health, finance, heavy manufacturing, transportation, aviation, e-service and e-commerce, military , music and movie, toys and games entertainment etc. industries.

In fact, many ideas, systems and technologies have been developing in the world of (AI) technology. However, which are net called or considered (AI) products, rather which are mentioned with their specific names, such as smart graphics, machine learning, e-commerce etc. (i.e. this is called (AI) effect).

What is influence to (AI) job nature change to change developed and countries countries digital economy
● How can (AI) technology influence digital economy?
Nowadays, (AI) related industrial applications will replace most human power in fields, including call centers, customer services and air cargo transportation. (AI) technologies also help weather forecasting based on repeated rainfall pattern (data) recognition, through robotics (i.e. floor cleaning, moving lawns etc.) transporting people and products with unmanned vehicles, sending space unmanned smart shuttles, developing robotic arms, predicting market values in stock exchanges by internet, making homes safer, helping elderly and disabled using robotic servants etc.

Among the (AI) related technologies , there are a few that significance for the impact on society and especially on digital economy . (AI) is particularly influential in machine learning. Such as robotics, transportation, finance, health and bioinformatics, e-commerce , e-games, big online data gathering and internet-of-things. For example, machine e-learning is based in bioinformatics and robots that can learn new skills for better caregiving in healthcare. What is machine e-learning? Machines can e-learn from e-data gathering, coming up generalizations and making decisions to act in certain ways from internet.

There are important applications , such as e-machine perception, electronic online natural language learning processing, online search engines, online bioinformatics, online brain –computer interface, online game playing, online robot locomotion, online advertising, online computations finances, online health monitoring, online DNA classification and decision making, online in chemistry –cheminformatics . So, online machine learning can positively impact productivity and it can enhance information and analytical system from (AI) online channel.

What is robotics? Robotics is one of the most strongly influenced fields in (AI). For example, heavy manufacturing industries, robots and used and man power is replaced for effectiveness, precision, and accuracy, especially in respective or dangerous tasks, including welding, assembling , picking and placing .

So, robots can acquire new skills or adapt the changing dynamic environment. Also, artificial intelligence can be applied in developing transportation. For example, automated vehicles, driver assistance systems , safety systems, collision avoidance systems and public transportation. Moreover, (AI) technology has proven to produce some of the best tools to predict stock market fluctuations from internet data gathering method. It's predictions are based on ever-evolving predictions algorithms and systems learn new models and make connections between historical data and new data to measure stock market trading more accurate from internet data gathering channel.

In health field, especially in health data processing , analysis, decision making support and medical diagnosis. So, online data can show which patients will need what treatment and what alternative drugs could be used more accurate from (AI) online data gathering method. Bioinformatics is an interdisciplinary field combining statistics, (AI) online technology can help in discovering data patterns and modeling through the application of machine learning, artificial neural networks and genetic algorithms. For example, further (AI) technology development of human genome project of online data sequences.

Online shopping can be facilitated by virtual assistants developed through (AI) technology and these assistants can offer the best advice. (AI) online purchase coming after every product image recommendations and personalization bring important revenue to shopping online sites, like Amazon . Smart computer graphics and games, artificial intelligence is useful in smarter computer, graphics, scene modeling , scene rendering processes in order to create, for example, effective human –robot interactions , online machine learning, online strategic games techniques etc. online computer related (AI) software.

So, online big data analysis and big data does have a critical need in the world of online intelligence machines and software in our future. In other words, (AI) offers online technology to enable online big data analysis to provide industrial organizations with valuable information for effective decision making in short time. For example, what IBM's Watson achieved: this machine used 200 million of structured and unstructured content with a special technology of hypothesis generation, massive evidence gathering, analysis and scoring from internet channel.

Finally, (AI) online technology another related internet invention (internet of things) (IOT) is the network of machines or objects connected through internet. These connected objects can sense their internal and external environment, communicate with each other, can send critical data and finally can make decisions to act or correct their environment from (AI) online technology. For example, factories can monitor and automatically change production processes, hospitals can monitor and regulate the health conditions of their patients , schools can collect data from facilities and cars can send data to car makers from (AI) online technology.

Partner predicts that (IOT) market will create about trillion amount value by 2020 year. Although machines collect big data from their environment, whether which gain an insight or learn from these online data largely depends on the (AI) online machine learning principals and (AI) online technology. In 2013, Mckinsey estimated that disruptive technologies closely related with potential economic impact in 2025 year between $7.1 to $13.1 trillion amount (automation of knowledge work, advanced robotics, autonomous or near-autonomous vehicles).

What is the relationship between
(AI) and global digital economy development ?

● Could work activities in China be automated
making in the nation with the world's largest automation potential?
Can (AI) technology influence China economy? Could China workers be affected and jobs made up of routine work activities and predictable? Will programmable tasks be particularly impact to China employment market ? When impact on labor market is likely to be gradual at the aggregate level, it can be sudden and dramatic at the level of specific work activities, rending some job obsolete fairly. Overall (AI) technology will raise digital skills when reducing demand for medium incomer inequality for China workers. It seems (AI) technology's effect on productivity could be crucial to China's future economic growth as the population ages are increasing.

In China, some biggest technological companies driving significant investments in research and development. Moreover, China is one of the leading global (AI) technology development county. However, China will need to focus on building its innovation capacity. For example, United States and United Kingdom are currently producing more influential (AI) technological research. However, if China planed to achieve (AI) technology success, it's traditional industries will need to develop technical know-how –to and overcoming implementation costs prepare to develop (AI) . When (AI) technology is introduced into China society, China government needs to raise concerning ethical, legal, technological security etc. business questions. Also, surrounding issues include privacy, discrimination, legal liability and regulation. It aims to encourage overseas investors to choose to invest (AI) technological industry to raise GDP growth and manufacturing industries income growth for long term in China. If China encouraged overseas (AI) technology investment in its country. It is possible to influence China employment market to be changed. Because (AI) technology will impact to influence China people daily life. Due to (AI) technology is introduced to China society, many rich people will prefer to spend to buy any high (AI) technological products for entertainment or learning or machine man driving etc. daily necessity activities. Then it will raise GDP growth and will raise (AI) manufacturers or related-(AI) technological manufacturers profit. It is beneficial to China because it can become one high knowledgeable and (AI) technological economical society. But it will bring bad influences to raise unemployment chance for the low skillful labor. In labor economy aspect influence , how (AI) technology can influence China low skillful labor unemployment ratio raising. The raising low skill labor unemployment reason is because China low skillful human labors are argued or are replaced by (AI) technology creating new challenges to introduce to influence China society of simply human manufacturing job nature to be changed to be high (AI) technology manufacturing job nature in any China factories. Moreover, when (AI) technology introduction to China, it will cause other related social challenges in China. The varied (AI) related challenges, including the difficulty of creating safe and reliable hardware for sensing and affecting (transportation and education), the challenges of gaining public trust, a low resource comities and public safety and security, the challenges of overcoming fears or marginalizing humans in China employment and workplace and the risk of diminishing interpersonal trust because the low skillful labors won't believe any China employers will give chance to employ them , due to (AI) technology will replace their skills and man manufacturing of productivity is much less to compare to (AI) technology manufacturing method.

● How does (AI) technology influence
the future of employment change?

Are future nature of jobs changed to computerization from (AI) technology? Where are the probability of computing occupations from (AI) technology influence? What is expected impacts of future computing on labor market from (AI) technology influence? John Maynard Keynes's frequently cited prediction of widespread technological unemployment " du to our discovery of means of economic the use of labor outrunning the pace of which we can find new used of labor" (Keynes, 1933, p.3).

In the future, (AI) technology will impact some nature of occupations to change computing. This chance will also influence some countries' economic change. For example, some factory human labors hand routine manufacturing tasks will be changed to computerization of routine manufacturing tasks by (AI) technological machine men hand manufacturing method. it will cause a structured shift in the labor market, with workers reallocating their labor supply from middle-income manufacturing to low-income service occupations.

Arguably, this is because the manual tasks of service occupations are less computerization, as who require a higher degree of flexibility and physical adaptability. So, (AI) technology will influence the human hand labor skillful occupation nature of task cheaper , such as vehicle manufacturing , ship manufacturing, computer manufacturing, steel manufacturing, television, radio etc. home electronic products of heavy machine industry change. Due to (AI) technology machine man will be proper to be used to manufacturing these electronic products when the (AI) technology innovation can develop to the mature stage. Then, any countries manufacturers will choose to use (AI) technology machine man, instead of human hand production.

Supposing the future prices of computing are fallen, seriously, problem solving skills are becoming relatively productive, explaining the substantial employment growth in manufacturing occupations, involving cognitive tasks where skilled labor has a comparative advantage, as well as the increase education needs for (AI) technology computing of machine man subject study.

Prediction of education needs for (AI) technology student numbers will increase, due to manufacturing industry needs many (AI) technology students in future employment market. Another (AI) technology influence if the future (AI) technological innovation, e.g. machine man manufacturing or machine man service industries will both increase demand, then with more sophistic software technologies will be disrupted labor markets by marketing workers redundant.

For publishing industry, what is striking about the case in paper book publishing industry will be unpopular? Due to the electronic book publishing industry will be popular, e.g. Amazon publish . (AI) technology can influence paper book manufacturing method which is replaced by machine man electronic book manufacturing method as well as it will cause the computerization is no longer confined to routine manufacturing tasks. Due to (AI) machine man manufacturing technology will be proper to be used to manufacture any products in short time efficiently and effectively , e.g. electronic book products. In the future, if it is fact to occur this case, such as (AI) technological machine man manufacturing method will be adopted (applied) to manufacture electronic books or any products in possible. (AI) technology will cause many manufacturing workers are unemployed. It is beneficial to employers, who can reduce to spend much wages expenditure to employ manufacturing workers, but it will cause many manufacturing workers loss jobs and reduce income to support whose families lives. It will cause social challenges, e.g. increasing stealing crimes if the manufacturing workers had not other skills to find other jobs to do easily. So, manufacturers need to concern over technological unemployment which will be hardly future phenomenon if who decided to dismiss all manufacturing workers, duc to (AI) tcchnology machine men replace to them.

If (AI) technology can be innovated to produce any kinds of machine man to serve any service or manufacturing industries successfully. Then, it will bring these questions: Can future that workers be influenced to be automation employment and productivity by (AI) technology influence? Does it impact to influence the (AI) technology countries' productivity and growth and natural resources development and labor markets and evolution of global financial markets and economic impact of technology and innovation and urbanization etc. issues? How will automation transform the workplace? What will be the implication for employment? What is likely to be its impact both on productivity in the global economy and on employment?

In fact, automatic of activities can enable businesses to improve performance by reducing errors chance and improving quality and speed, and same cases achieving outcomes that go beyond human capabilities. Some economists indicate (AI) technology would give a needed boost to economic growth and prosperity have of the working age population in many countries. Based on the scenario modeling, they estimate automation could raise productivity growth globally by 0.8 to 1.4 % annually. They also indicated that almost half the activities people are almost $1.6 trillion in wages to do in the global economy have the potential to be automated adapting current demonstrates technology, according to their analysis of more than 2,000 work activities across 800 occupations. When less than 5% of all occupations can be automated entirely using demonstrated technology, about 60% of all occupations have at least 30% of worker made activities, that would be automated. More occupation will change to be automated. They also indicated for business performance benefits of automation are relatively clear, but the issues are more complicated by policy making to attract foreign investors. Beyond technical feasibility, the cost of technology, competition labor will include skills and supply and demand dynamics, performance benefits and beyond labor cost savings and social and regulatory acceptance will affect the automation. Their predictions suggest that half of today work activities could be automated by 2055 year, but this could happen 10 to 20 years earlier or latter depending on the various factors in addition to their wider economic condition.

Some scientists suggest (AI) technology is finally starting to deliver real-life business benefits. Computer power is growing significantly , algorithms are becoming more sophisticated and perhaps most important of all, the world is generating vast quantities of the fuel that powers (AI) technology data billions of gigabytes of it every day. Also, online firms are digital natives, such as Google online search service company is investing on (AI) technology. For new though most of the news if coming from the suppliers of (AI) technologies. And many new users are only in the experimental phase. Few products are on the market or are likely to arrive these soon to drive immediate and widespread adoption. As a result, analysts believe (AI) technology's potential will give true economic benefit in the future. (AI) industry will introduce to suppliers and users to raise economic potential of (AI) technology.

In the future, (AI) technology systems can solve business problems. Some scientists categorized those into five technology systems that are key areas of (AI) technology development: robotics and autonomous vehicles, computer vision language virtual agents and machine learning , which is based on algorithms that learn from data without replying on rules-based programming in order to draw conclusions or direct an action.

Such as computer vision and language includes natural language processing, analytics, speech recognition technology, some are about learning from information, such as about machine learning and others are related to acting on information, such as robotics, autonomous vehicles and virtual agents, which are computer programs that can converse with humans. Machine learning and a subfield called deep learning are artificial intelligence applications.

- ● Can artificial intelligence impact

global economy growth?

Artificial intelligence (AI) is a term first defined in 1956 year. It is a branch of computer science that aims to create intelligent machines that work and react like humans. In contrast today, 60 years later, (AI) is characterized by a number of applications, including computers playing games against humans and understanding human languages, virtual personal assistants, and robotics which involve computers seeing , hearing and reacting to sensory stimuli. In the future, technologists predict for (AI) technology ranging from (AI) being used as a tool to aid relatively simple processes for robots with human like mental capabilities, who expect (AI) technology can emulate human performance by learning, coming to mind its own conclusions, understanding complex content, engaging in dialog with people, enhancing human cognitive performance or replacing humans in executing both routine and non-routine tasks. In existing industry, (AI) technology is used , such as targeted advertising and virtual used personal assistant as well as the (AI) technology that my exist in the future, such as robots with human vehicle processing capabilities.

The range of (AI) technology's progress in the future will determine the economic impact future of (AI) technology on the global economy with more limited advances and applications (i.e. weak (AI) only) corresponding to more limited economic impacts and more substantial progress, i.e. strong (AI) technology is corresponding to more significant economic impact.

(AI) technology learning that automates analytical model, including predicting cause-and-effect relationship from biological data, identifying new drugs, self-driving cars and protecting against fraud etc. functions. Also (AI) learning can improve natural language processing that allows computers to continue to better analyze, understand and generate language to interface with human using the natural human language, virtual personal assistant, helps users by providing scheduling appointment, reminds organizing personal finance and finding providers of various services, machine vision allows (AI) machine man to identify object, scenes and activities in detect pedestrians and bicyclists.

We expect the economic effects of (AI) technology to include both direct GDP growth from sectors that develop or manufacture (AI) technology and indirect GDP growth through increased productivity in existing sectors that employ some from of (AI) technology. If (AI) producing sectors could grow, then it could lead to increase revenues and employment of (AI) technological professionals within these existing firms as well as the potential creation of entirely new economic activities to any countries' societies productivity improvement in existing sectors could be realized through faster and move efficient processes and decision making as well as increased (AI) technological knowledge and access to information available in societies easily.

In the future, if (AI) technology is an increasingly critical component of more products, it will become an integral part of necessary products of many people's lives. The extent of (AI)'s economy effort is also likely to vary from region to region, thought variation may be more dependent on the predominate economic activity of a region and the (AI) ability can influence economic activity, rather then the economic or developmental status of the regions. (AI) technology can move accessibility and can use source development to do international business between one country and another country.

So (AI) technology has the potential to give benefits to different income chooses and to bring significant gains to both developed and developing countries. For agricultural technology, (AI) has the potential to optimize food production around the world by analyzing agricultural regions and identifying what is necessary to improve crop yield. In total, (AI) technology gives greater economic impact to any countries agricultural regions if which implemented (AI) technology to grow crop , fruit etc. food production in the farms.

Investment in (AI) technology is such as capital investment to any countries' public or private enterprises. So, it will have large economic impact to the future . If the (AI) technology is reasonable invested to the different needs aspect by the public or private enterprises in the country. Then, it will have good economic impact to the country in the future. However, when (AI) technology is likely to affect both the productivity and employment components of economic growth in many sectors. Significant public debate has focused on projections of (AI)'s effect on the labor force. However, for instance, some researchers have argued that the rise of (AI) technology and automation will led to significant unemployment as capital is substituted for the low skillful labor. So, they point to the concern that the increasing sophistication of (AI) technology may balance skilled and semi-skilled workers and the reduce the size of the middle class. However, this is not a new argument, due to (AI) technology negatively affecting the labor force and leading to mass unemployment. Because the (AI) technology is the substitution of machinery for human labor. Although, employment in certain industries, has been reduced in the past due to technological advancement. For long term, the labor market has adapted to the introduction of new technology, giving rise to new jobs in new areas. (AI) technology may also be accomplished without a reduction to total employment in the long-term to some Asia countries, such as Hong Kong and Japan. Because Hong Kong and Japan many low skilled labor, e.g. security, cleaner who complaint that employers need them to work long time hours. (abnormal working hours) e.g. one day 12 to 15 working hour per day. Hence, if (AI) machine means invention technology success. Security or cleaning job can be worked from (AI) machine man in some hours every day in order to reduce the long time working hours cleaners or security workers, e.g. one (AI) machine man works 4 hours for cleaning or security job, one day as well as another cleaner or security labor only needs to work 8 hours one day. So total security or cleaning employers can employ 12 hours machine cleaners or security workers and human cleaners or security workers in one day. For long term benefit, Hong Kong or Japan every security or cleaning worker does not need to work 12 hours minimum working hours one day. They won't feel tried and bore and without private with whose families, so who will accept to do these cleaning or security jobs, even they can raise work efficient and performance when who feel happy and health.

So, (AI) technology of machine man invention can raise low skillful labor efficiency and it can help them to avoid abnormal working hours demand in some busy work life countries, such as Hong Kong and Japan. Before, one Japan female labor feel unhappy to work, due to who often needs to work abnormal working hours for her employer and who has less sleeping and without any private time to enjoy her life with her families every day. So this abnormal working hours factor causes her to do commit suicide behavior, then she is die unlucky. So (AI) technology of machine man invention ought avoid abnormal working hours demand for employer in any countries in the future.

The most important occurrence to any employers, some researchers had attempted to do one experiment to find that private research and development , venture capital and public research and development investment all have strong net effect or economic growth with venture capital funding further having the strongest such effect from (AI) technology. The researchers hypothesize the venture capital investment contributes to economic growth through (AI) technology innovation and by the capacity of an economy to use existing (AI) technology knowledge to increase productivity. They predict the impacts of venture capital, business-research and development and public research and development can raise multi factor productivity from (AI) technology introduction.

Can (AI) technology influence the economic development to developing countries? The developing regions of the world contain most of natural resources. If one day, (AI) technology has invent one kind of machine man which can assist any gas or oil workers to seek any new oil/gas natural resource locations easily. I believe that (AI) technology can help these natural resource exploitation countries will gain economic benefit more easily. So, (AI) driven technology can be used to change to create any new opportunities to address poor management or resources and improve human well being, such as Africa Latin America and India can use (AI) technology machine man to seek any oil/gas natural resource countries exploitation activities to attempt to gain much economic benefits.

● Why will (AI) technology grow economic development ?

Nowadays, increases in capital and labor are no longer driving the levels of economic growth, such as (AI) technology. The ability of increase in capital investment and in labor of traditional drivers of production, have no longer to be enjoyed in most developed economies ,e.g. developed country, US, UK . However, artificial intelligence has the potential to overcome the physical limitation of capital and labor to avoid missing out on this opportunity. So, policy makers and business leaders must prepare for and work toward a future with artificial intelligence. They must do with the idea that (AI) is another simply method to enhance productivity method . Rather they must see (AI) as the tool that can transform thinking about how growth is created.

Economists have always thought of new technologies are as driving growth their ability to enhancing. It can replace labor and capital factor of production. So, it brings this question: What is the factor of production (AI) technology characteristics. They key factor is to see (AI) technology as a capital-labor .

(AI) can replicate labor activities at much greater scale and speed, and to even perform some tasks began the capabilities of human. For example, by using virtual assistants , 1000 legal documents can be reviewed in a matter of days instead of taking three people six moths to complete. Some (AI) technology may be one kind of factor of production in the future. For another example, people will work in workplace digitalization environment. So, in the future, working environment and information management are automated. Such as Konica camera sale company will use workplace digitalization. So , (AI) technology can provide workplace digitalization in order to raise productivity efficiency. (AI) technology will be one kind of production which is replaced by workplace digitalization and it will grow any organization productivity efficiently. Then, (AI) technology will assist overall social economy growth , due to productivity is raised and products can be produced in short time to prepare to sell in consumption market. So, time will be shortened to increase GDP growth fast for the development of (AI) technology countries.

● How can (AI) technology impact to global
economic and social and psychological
changes?

What will be the development of (AI) technology and predictions concerning the future evolution? The computers and robots will develop conscious, intelligent and minds into humans, enhancing psychological and behavioral abilities and allowing for direct communication with (AI) minds. (AI) technology will be impacted human life by (AI) technology information communicative and environmental influence. A " world brain" and " world mind", this psychological system will be enhanced and enriched the capacities of both individual and collective cognition by (AI) technology of service industries.

(AI) technology with influence these human needs of service industries changes, such as , biological science, finance, entertainment, business, biological science, transportation, communication military etc. The personal computer evolution, the internet and the world wide web which exploded on the scene, linking business, homes, schools, social organizations which were a completely unpredicted phenomenon to influence human life. Kurzweil (1999) predicts that by 2029 year, most human communication will be with machines. According to Person, by 2100 year, there will be human machine convergence.

How can (AI) technology influence environmental protection to make benefits to farming economic growth? (AI) technology can be applied to predict how to solve environmental pollution challenge to avoid to damage any crop or vegetable or rice or fruit etc. food growth. Because environmental experts can gather global environmental pollution data from an environmental database to build a perform a systematic analysis from (AI) technology. The first step is this broad analysis can include understanding, statistical and data gathering techniques to obtain the relevant data, the correlation among the variables involved, and a list of possible models. The next step is to select a set of methods and models that cover all kinds of knowledge and functionalities needed for the decision making process. Once the models are selected, they must be fully implemented by means of machine learning , data mining, statistical or numerical technique. After that, those models must be integrated to build the whole EDSS. The EDSS must be tested to check its performance, accuracy, usefulness and reliability, both from the user's and (AI) technology/computer scientist's point of view. If these is any wrong feature in any development stage, such as model's integration, models' implementation, selection of models, database, problem analysis etc. the developers must come back in the update th required components. When the evaluation phase is all right, the EDSS is ready to be applied to the environment. The great contribution of artificial intelligence to EDSS the integration of several methods complementing the classical statistical models/simulation , statistical analysis, linear models, etc. and numerical models (control algorithms, optimization techniques etc.) .

This cooperation makes the resulting systems more reliable and powerful in coping with real world environment systems. Date interpretation has been a principal area of research in (AI) technology since the very beginning. The most demanding problem in the environmental assessment context.

Knowledge representation permits the definition of the different types of data that the existing methods adapt to the process. There is also a lot of work to clean, repair and transform the huge available quantities of raw data. Apart from this, the availability of meta-information or background knowledge is required to guide the process. Data mining is multi-disciplinary: It covers expert systems, data based technology, statistics, data visualization and unsupervised machine learning. These techniques operate at the level of data and background information, where numerous and often incompatible new commensurate pieces of information from disparate sources have to be brought together (K, Fedra, 1994).

So, it seems that in the future, (AI) technology with the increasing maturity in particular those related to knowledge and engineering, new dimensions can be assisted to users in environmental decision making are available. For example, many environmental systems are characterized both by incomplete models and by limited data. Hence, in the future, (AI) technology will be applied to predict climate change to reduce crop or fruit etc. food agriculture challenge by climate change bad influence.

● Will (AI) technology influence digital economy change to manufacturing industry ?

To understand how the manufacturing business must adapt to prosper in the technology, we need to understand how (AI) technology will change us to shape our daily habits to satisfy our expectation of products to how we shop and even the immediate of the entire process. For example, taxi services are in the crosshairs as on demand transportation services like, available of the touch of a smart phone button expand. In fact, Yellow lab, US country , san Francisco city's largest taxi company is filing for bankruptcy as the industry starts to change faster than almost anyone expected. However, at this point, its more than an app that is changing, some our taxi passengers renting taxi transportation to catch consumption behavior.

(AI) technology will influence digital economy for taxi passenger's individual customer experience, offering a growing renting taxi to catch of service and feedback opportunities when any one taxi passenger who chooses to use mobile phone app online tool to prepaid to rent any taxi more easily.

Also in the long term, (AI) technology can influence vehicles drive themselves of behavior. Already, companies like Google and GM are working on projects to bring fleets of autonomous vehicles to cities at the path of a button.

Moreover, this on-demand service model is beginning to appear across a much broader range of markets. For example , Amazon company is investing in its own fleet of trucks, planes and even drone at the same time as it pushes for same-day delivery of products. As some point, vehicles will be autonomous too. So, it seems that (AI) technique will influence any transportations choose to use digital autonomous driving technology in the future . For Amazon company case, it is not stopping of logistics. It is also aiming to automatically manage the supply of consumer home products with its recently launched Amazon replenishment service, Dash. Dash is a digital service that enables that connected derive to automatically order physical products from Amazon when supplies are running low. So, it seems (AI) technology will be applied to logistic function by digital technology method introduction in the future.

Hence autonomous vehicles will optimize industry supply chains and logistics operations through increased efficiency and flexibility. In fact, fully automated and lean supply chains will keep reduce load sizes and inventory by leveraging smart distribution technologies and smaller autonomous vehicles by machine man assistance. If Amazon continues to grow market share for online sales by reducing effort required by the consumer to place an order, when also contributing the almost immediate delivery of products to the doorstep. So, it will further fuel the trend toward on-demand derive. As Amazon company fuels the on-demand economy, consumers will expect immediacy in more parts of the digital economy. On top of speed, consumers increasing expect more personalization options.

So, (AI) technology will influence digital manufacturing, such as Amazon publishing to monitor every aspect of every process in real -time and communicating to self-optimized deep learning robotics, new methods of high volume and high customization will become possible. Then, as products merge into product platforms and even services, manufacturers have the opportunity to provide components and platforms used by smaller players. So, (AI) technology will influence manufacturing industry to choose automated SMI lines, robots installed, automation engineers.

Another future (AI) technology development can be applied to space science aspect, such as Automation engineering space in manufacturing process to achieve digital manufacturing benefits to any businesses in the future. Such as reducing cost, shortening manufacturing time, raising efficiency, shortening delivery products to client individual time. How can artificial intelligence give the need and advanced fast and evaluation methods benefits for space exploration? When US NASA (space exploration organization) achieves any space exploration missions, it will answer this question:

When is it useful to have a machine use (AI) technology to achieve a decision? After all, after millions of years of space exploration and rough 10,000 years of civilization, humans are usually quite good at making decisions in complex uncertain environments. Through, Johns Hoplains University's Applied Physical Lab. Research in (AI) technology enabled systems, which has identified three general use cases for (AI) technology to explore space mission:

First, for some tasks (AI) technology is more cost effectiveness than human. Second, (AI) technology is better suited than humans at solving some, but not all problems. Third, (AI) technology allows NASA organization's space exploration mission to develop machines that ate capable of responding faster than when a human is in the decision loop (D. Scheidt, 2012, A. Castano et. al. 2008).

So, the use of (AI) technology to enable science by observing the pace of rapidly evolving phenomena was demonstrated. It is more effectively coordinating and (AI) technology utilizing to earn economic benefits to use for space exploration mission.

However, (AI) technology also have current risk for space exploration. Today (AI) technology is immature and requires further development to reach its potential. For instance, the (AI) technology algorithms that detected the dust derive could not have identified whether the Martain weather represented a threat to the cover. Also it can not yet use instrument input to determine what, where and how to autonomously make the next space science measurement. An equally important factor limiting (AI)'s deployment is that lacks the methodology and technology to effectively test (AI) technology. So, the challenge will testing (AI) enabled system is how (AI) performance can be measured. It would be NASA organization's difficulty to find (AI) technology to develop to carry on researching any space exploration missions in the future. However, (AI) technology will be a good economic benefit choice for space exploration mission in the future.

● What is artificial intelligence potential
benefits and ethical considerations?
The ability of (AI) technology systems to transform vast amounts of

complex information into insight has the potential to help solve manufacturing or service challenges for human needs. However, to reap the societal benefits of (AI) systems, humans will need to trust then and make sure that which follow the same ethical principles, moral values, professional codes and social norms that we humans would follow in the same scenario, research and educational efforts as well as carefully designed regulation in order to achieve the most effort of economic benefits goals. For example, international business machines corporation (IBM) is actively engaged both competitors , in global discussions about how to make (AI) ethical and as beneficial as possible for people as social economic benefits.

(AI) is usually defined as the " capability of a computer program to perform tasks or reasoning processes " that human usually associate to intelligence in a human being. Often, it has to do with the ability to make a good decision, even when there is uncertainty, too much information to handle. As an example, play chess or complex card games of entertainment activities is believed to need some form of intelligence in a human being, as well as choosing the best medical facilities in a difficult medical case, or creating something new, such as mathematical theorem or even some form of act, or even driving automatic machine man (self driving vehicle) replacing human driving in the middle of a crowded city.

(AI) needs depends on what we consider being intelligence in the behavior of a human being act a certain point in time. If human belief about human intelligence changes and we don't believe any longer that a certain task requires intelligence, then a computer program performing that task is no longer part of (AI), it becomes just another boring computer program. So, it means that (AI) technology will replace some old computer programs, if human can invent new generation of (AI) software for any functions or activities to satisfy human needs.

As IBM, it argues intelligence. This means that we aim to build systems that enhance and scale human expertise and skills rather than replacing them. We therefore focus on practical applications of (AI) capabilities that assist people in performing well-defined tasks of needs by exploiting and wide range of (AI)-based services. We also use the term " cognitive computing" it is mean a comprehensive net of capabilities based on technology. It comprises the fields of machine learning, reasoning and decision technologies, language, speech and vision recognition and processing technologies, high performance and high efficient functions for any industries or individual consumers needs. For example, robotics, which are usually very good at doing what which are supposed to in any environment, much have public shopping center, factory etc. places which need simply services from the robot (machine man), such as cleans the floor of our houses to the robot that can work together with humans in production chains, passing through the warehouse, robots can take care of the tasks of an entire warehouse and the companion robots like Nao, Pepper, Aibo and Giraff, who can entertain use, talk to use and help elderly people to stay connected to their friends, relatives and doctors.

Google company is building automatic machine (self-driving cars) and has acquired more than 10 robotics companies. Facebook had opened whole new research facility only on (AI) research. Apply computer has developed Siri. Microsoft computer company has built a similar personalized assistant. Google has Deep mind, a UK company whose long term aim is to build general (AI) and has already great potential to win game to the world champion and IBM is investing a huge amount of resources in applying its Watson cognitive computing system to the medical domains to finance and to personalized education. In Europe, IBM is establishing new centers in Munich and Milan focused in the application of cognitive computer capabilities to the internet of things and healthcare respectively.

For example, automatic machine man (self-driving cars) are all about (AI), which used to be able to see what happens in the street (signals ,lanes, other cars, pedestrians, traffic lights, which need to able predict what other cars and pedestrians will do, and who need to be able to cope with unforeseen situations. Since, most car accidents are due to human fault, it is estimated that the adoption of self-driving cars will save about half of the lives that are usually last in car accidents.

IBM Watson company has to understand spoken language, make sense of massive amount to text , respond correctly to questions in many categories, as well as assess its own confidence in responding to such questions. In the future, (AI) technology can own question/answering capabilities that would be very useful, for example, in assisting a doctor when trying to some to the correct diagnosis for a patient and to propose the best therapy .

Intelligent machines can also rely on huge amounts of data to be used to learn how to make better decisions. This data comes from all of us over the years Facebook users have uploaded more than 250 billion pictures and every day who upload about 350 million more. Every second, we submit 40,000 google search queries. So, (AI) technology will be connected through the web from appliances to traffic lights from cars to watches.

Other tasks that are very easy for humans are physical and manipulation tasks, such as walking , running, picking up an object to make its shape and location, restricted environment. But (AI) machine man technology still not able to have the general physical and manipulation capabilities even of a 6 year old.

So, it brings this question: Why do (AI) scientists need to concern ethics? Because (AI) technology is complex, information into insight has the potential to reveal long held secrets and help solve some of the world's most difficult problems. (AI) systems can potentially be used to help discover insights to treat disease, predict the whether, and manage the global economy. So, ethic issues is important to and (AI) scientists . If any one new (AI) technology research investigation could success, it will be a secret to and the (AI) scientists can not permit to their loyalty to any competitors to damage the fair (AI) technology products trading market. The country (countries) (AI) technology scientists need to concern ethic issues, who need to keep secrets for their countries economic or/and social benefits. This is moral issues to any countries/country loyalty is whose countries intangible assets. They can not sell (AI) loyalty to any their countries to assist whose economic benefits immorally.

● How can (AI) technology influence to global
health care economy development?

According to (AI) lecturer analysis, when combined key clinical health (AI) application can potentially create $150 billion in annual savings for the US healthcare economy by 2026 year. (AI) technology is re-winning modern conception of healthcare delivery. It enables machines to sense, comprehend, act and learn. So which can perform administrative and clinical healthcare functions (Accenture, 2017).

It will help health care service organizations to reduce health care cost, will improve and raise service quality and access. So, (AI) health market size will be predicted growth. (AI) applications in health care include robot-assisted surgery, virtual nursing assistant, administrative workflow assistant, fraud detection, error reduction connected machines, clinical trial participant identifier, preliminary diagnosis, automated image diagnosis and cybersecurity.

What kind of benefits (AI) technology can contribute to healthcare service? (AI) technology can deliver what many health care organizations need, such as financial and operational of labor costs, digital expectations from patient consumers how to use (AI) technology to solve interoperability challenges in any healthcare organizations. Also (AI) technology can be applied to wellness an d lifestyle management, diagnostics, delivers financially but also way of organizational and workflow improvement. So, (AI) technology will be continue to become most prevalent and adoption to healthcare organizations , which must need to enhance structure to be position to take full advantages of new (AI) technological capabilities. (AI) technology can change the nature of work and employment is rapidly changing to make the best use of both humans and (AI) talent in healthcare industry in the future. For example, (AI) technology offers a way to fill in gaps and the rising labor shortage in healthcare. According to Accenture analysis, the physicians shortage is increasing. However, (AI) technology will manufacture healthcare machine men to replace physicians in future one day(2017). Hence, (AI) technology will be invented to raise health care service staffs work efficiency and performance in any hospitals or clinics in the future.

In conclusion, (AI) technology will raise efficiency for any service or manufacturing industries in the future, although, it is possible that it will also rise low skillful workers unemployment numbers. But, the most important influence to human technological innovation will be risen and it will influence human life will be changed to be better, e.g. self drive cars, health care physician machine men, machine man cleaners etc. intelligent machine men will be manufactured to serve for our daily life.

machine men will be manufactured to serve for our daily life. Furthermore, (AI) technological products will influence countries trading, some low technological development countries manufacturing businessmen can choose to buy any (AI) products to raise whose productivity and efficiency and reducing cost to achieve economic cost saving result. Also, GDP of trading growth income will increase to the (AI) products sale countries. Hence, it will be beneficial to economic development to both developed and developing countries both in the future as well as (AI) scientists time and money spending will be valued to continue to invest (AI) technology development for human life and economy benefits for long term.

In conclusion (AI) technology will raise macro economy growth and it can create many (AI) jobs , but it also raise the low level technological worker unemployment change. In the future, (AI) technology can be applied to digital technology to attempt to invent any new undiscovered (AI) and digital technology. So, it needs any scientists to continue to research how digital and (AI) technology can be mixed to satisfy human's future undiscovered needs.

Must Developed And Developing Countries Need Artificial Intelligent To Replace Human Job

Must developed and developing countries need artificial intelligent development? If one developed country, e.g. US, UK , Japan , Singapore it does not continue to develop artificial intelligence, robotic, then what disadvantges or weaknesses , it will encounter to compare when it chooses to continue to develop this artificial intelligent technology in society. If one developing country, e.g. China, Korea, Taiwan, it does not continue to develop artificial intelligence, robotic, then what disadantages or weaknesses, it will also encounter to compare when it chooses to continue to develop this artificial intelligent technology in in society. I shall explan the reasons why the results may cause to either the developed country, or the developing country as below:

● How AI help developing countries to communication and agriculture and learning and medical delivery development

Why can AI help developing countries ? Drones that pick inaccessible crops and mobile phones that give medical advice are two of the ways AI can transform life in the developing world. Artificial intelligence (AI) may improve the lives of the world's poor, the technology needed to revolutionise inefficient, ineffective food and healthcare systems in developing countries is well. For example, in low-income areas, agriculture and healthcare are two critical ecosystems that we can apply AI to immediately; this is not the far future, or even in five years.

Artificial intelligence (AI) has seeped into the daily lives of people in the developed world. From virtual assistants to recommendation engines, AI is in the news, our homes and offices. There is a lot of potential in terms of AI usage, especially in humanitarian areas. The impact could have a multiplier effect in developing countries, where resources are limited.

Emergency Response to developing countries' earthquake natural damage suddence occurrence predicting

AI and machine learning are still finding importance in emerging markets, but certain applications have emerged and are now widely used. For instance, predictive models for disaster relief enable first responders to automatically analyze large-scale behavior and movement through multiple sources of data including social media platforms, web forums, news sources, etc. Based on collected data, responders can scale reconstruction efforts and distribute supplies in a timely manner.

Why and how AI can assist farmers to predict when the earthquake occurs suddenly in order to avoid or reduce the natural damage to their agriculture productive number loss. For example, In 2015, when a major earthquake hit Nepal, more than 8 million people were affected. During the aftermath, drones were used to map and assess the destruction and speed up the rescue mission. The town of Sankhu, situated about 20 kilometers northeast of Kathmandu, was among the highly affected locations. In May 2018, my company Fusemachines and GeoSpatial Systems partnered with Sankhu's city officials to use drones and artificial intelligence in an effort to automatically estimate the reconstruction need. After processing data accumulated from a drone-powered aerial mapping of the region, the team fed this data to advanced machine learning algorithms. Combining drone imagery, digital mapping and machine learning, the team configured region modeling and infrastructure development with higher accuracy. Another organization known as One Concern, a California-based startup, has created a predictive AI program called Seismic Concern to accurately predict seism and is also working on solutions for wildfires, floods and hurricanes.

Smart AI Agriculture

Another application of AI in developing countries is smart agriculture. Farmers monitor crops more effectively and make better predictions on planting, weeding and harvesting using AI tools. It can also be used to analyze one plant at a time and add pesticides only to infected plants and trees instead of spraying pesticides across large swaths of crops. One California-based tech company is an example of this use of AI. So, the developing countries farmers in rural parts of India are also using AI to increase yields through better access to information about the farming season than they would normally have. Technology-enabled process automation offers the agribusiness industry the chance for remarkable growth -- not only in developed countries but around the world. There's a unique opportunity to increase yields, cut down labor costs and improve people's health.

Medicine Delivery to developing countries' patients urgent need

Companies are also leveraging AI to improve access to health care in some of the most remote areas of the world. In Rwanda, for example, Zipline is using drones to deliver medical supplies and blood to hospitals and clinics that are difficult to access by car. This has dramatically impacted people living in remote parts of the country because they are able to get medical help when needed. The drone system in Rwanda has also helped reduce waste of blood by 95%, as noted by Zipline. One Concern has created an AI program called Seismic Concern that accurately predicts seismic events and is also working on solutions for floods, wildfires and hurricanes. The medical field may actually benefit the most from emerging technologies in developing countries.

Assistance to reduce teaching work workload or psychological pressure to teachers in developing countries' schools

Another vital area benefiting from innovative technologies like AI is education. Advanced technologies can enhance how we learn, teach and perform tasks. In most developing countries, schools lack experienced teachers and resources to enhance students' knowledge. As a result, many students still have to walk long distances to get to the nearest school, which has created education gaps, especially in rural areas. AI tools such as personalized learning assistants can simplify learning by making tutoring services and learning materials accessible to all students, wherever they are. Machines can be automated to help students learn basic concepts without a tutor, which companies like Carnegie Learning are working on. This would allow students to learn at any time from anywhere. With AI, education is made easy and accessible to more people.

The initial usage of AI in developing countries has been at a micro level -- solving small, specific problems in a defined industry. As machine learning advances and there is a higher utilization of AI, we will see more complex issues being targeted and resolved. When duly adopted, AI can positively impact future developing countries people everyday lives not just in disaster intervention, education, health care and agriculture but can also help in mitigating poverty, malnutrition and pollution. Especially, in developing nations, to leverage AI's true potential and create a snowball effect. Startups are defining a holistic and humanitarian approach to building more sophisticated, AI-ready societies. Stakeholders in the AI landscape should understand the strengths and nuances of the developing world as well as the limitations of AI and create localized solutions and applications.

Why does smart phone help developing countries communication ?
Internet Seen as Positive Influence on Education but Negative on Morality in Emerging and Developing Nations. Internet access differs substantially across the 32 emerging and developing countries polled, with the lowest rates of internet use in South Asian and sub-Saharan African nations. Within countries, computer owners, young people, the well-educated, the

wealthy and those with English language ability are much more likely to access the internet than their counterparts. To access the internet, people increasingly use smartphones rather than more cumbersome fixed landline connections and computers. Around the world, both smartphones and basic-feature phones alike are used for sending messages and taking pictures.

In fact, many developing countries young people, students are popular to use smart phones for internet usage aim, instead of communication. Moreover, many developing countries working people are also popular to use smart phones for any working usage in their working time , even non working time any time. So, smart phones (AI) phones will be important communication or leisure tools to developing countries people in the future. Unless, it is one day, scientists can develop another new communication tool to replace smart phones. So, artificial intelligence will be important to influence developing countries people , how to improve or bring positive learning attitudes to students in their daily learnnng lifes. as well as how to raise developing countries people, how to raise working people efficiency or improve performace in their daily working lifes. So, AI may bring positive learning or working attitudes to developing countries working people and students both.

The Positive Impact of Mass Media in Developing Countries
Radio, newspapers, television, Internet, social media, etc., all of these are forms of mass media. Each of these outlets has the capability of bringing information to thousands of people with one device. While in some communities it is easy to take advantage of these communication outlets such as television and Internet access, not everyone has access to such outlets. Radio is one of the most common forms of mass media in developing countries because it's affordable and uses less electricity than many other forms of mass media, but only approximately 75 percent of people in developing countries have access to a radio, and roughly 77 percent of people in rural areas have access to electricity.

For developing countries that have implemented forms of mass media in their communities, there have been numerous positive outcomes are influenced to impact developing countries mass media by artificial intelligence as below:

When AI is participated to developing countries mass media, it can influence any radio, television audiences raise more attention to each other through social media platforms such as Facebook and Twitter and create, organize and initiate street protests and campaigns. Furthermore, having access to social media in developing countries, people are able to connect to those that they usually wouldn't have the chance to talk to. Moreover, AI Provides educational opportunities- In many countries, the division between local and national languages as well as issues of literacy can make communication difficult. With the use of mass media, a bridge can be built between these two gaps. In India, there is a radio station that provides information in local languages and respects local culture and traditions. One of the main ways is to create public awareness of what is going on with businesses and government officials. The media plays an important role in giving people the opportunity to act against injustice, oppression and misdeeds that they otherwise wouldn't know about. Information on available healthcare, a mass radio broadcast was sent out encouraging parents to seek treatment at local healthcare facilities for their sick children. With this mass outreach on healthcare, the encouragement of people to take their children to healthcare facilities saved thousands of lives. This easy way of encouraging others and bringing awareness about certain diseases was made possible through a simple radio broadcast. Finally, when AI is particiapted to media, it may bring many social issues to life that otherwise would remain unknown to many people. In developing countries and communities like Burkina Faso, when the radio broadcast was released about malaria, diarrhea and pneumonia, people were educated and moved to action and knew to take their children to healthcare facilities for preventative care. As it is seen, having access to different media outlets is vital for those in developing countries. Here are three ways that those in developing countries can implement mass media to help their people and communities.

When AI is participated to any internet radio or internet newspaper mass online listening or reading channel. It can provide online radios or newspapers in public places- By providing online radios and newspapers in public areas it gives community members to access news, information and emergency warnings. Even though radios can be on the cheaper side, there are still many people that can't afford to have a radio in their home. By providing one in a local place, not only would it better educate the community members but also it will bring the community together. So, it can make media outlets a two-way platform- Creating a two-way platform between the community and those who are behind the radio stations, newspapers or broadcasts makes the community feel involved and that their voices are being heard. An organization called Soul City in sub-Saharan Africa is showing how well two-way platforms work by engaging their listeners and having them contribute thoughts and ideas about complex issues. Because developing countries radio listening audiences or newspaper readers are popular to accept computer online radio listening channel or online newspaper reading channel to replace traditional paper newspapers or radio machines. So, AI may raise their listening news or reading news leisure feeling from online mass media channel in the future.

● Why do developed countries need to develop AI

Artificial intelligence, or AI, is driving massive shifts across the globe, and every day more questions arise. What impact will AI have on the workforce and how can we prepare for it? How can we encourage economy-boosting and job-creating technologies? How can we ensure that AI will be implemented ethically and with minimal bias? How will society benefit? For developed country, such as US example. None of the US, Israel and Russia have a formal national AI policy yet. Private sector companies such as Google, Amazon and Apple and the US department of defence are driving the bulk of AI investment in the United States. Though Israel does not have a specific policy, it is keenly focused on AI and has seen the number of AI start-ups triple since 2014.

Developed country may learn whether what weakness it is lacking when it does not continue to develop AI from one another developed country. Which countries are approaching AI most effectively, and to what degree is there opportunity for greater international collaboration? It may be too early to tell; however, when analyzing the best practices of existing national AI policies, there is much that can be learned. These are the specific areas to consider. When one developed country continue to develop or research AI, it may bring these benefits as below:

On gathering Data aspect, from self-driving vehicles to smart cities, data is the driver behind AI. Innovation in the United States is limited without a national strategy that answers questions about protocol and ownership. France and Denmark, on the other hand, are opening government data. France is hosting troves of centrally collected public and private data that it plans to make available as part of its strategy. Conversely, by taking a restrictive position on issues of data collection (as indicated by the implementation of General Data Protection Regulation), the EU is putting manufacturers and software designers at a disadvantage while balancing the demand for privacy. On raising technologica talent aspect, the demand for AI talent far outweighs the available supply. As a result, almost every nation's strategy addresses talent development. Canada's AI strategy is distinct in that it primarily focuses on research and talent strategy. The country boasts AI degree programmes and is building a $127 million research facility in Toronto. Companies like Facebook and my own company, Uptake, are investing in Canada to access this talent pool. On AI legal technological innovation aspect, a whole host of legal questions swirl around AI. The country is developing a bill for AI liability that will be ready in March 2019. The government hopes the legal framework will attract investors by providing a simple, comprehensive guideline to enable the broad use of AI systems. So, when the developed country applied AI technology to assist any lawyers to work, then AI can help them to reduce the workload to draft any legal documents more easier. So, any developed countries lawyers' draft legal documents time must reduce if the developed countries lawyers accept to apply AI to assist their legal works. One of the great promises of AI is its potential for improving quality of life. But without the right planning and oversight, we risk exacerbating problems of inequality or marginalizing groups of people. As an example, India's AI strategy is focused on leveraging the technology not only for economic growth, but also for social inclusion.

AI may bring what benefits to developed countries
From SIRI to self-driving cars, artificial intelligence (AI) is progressing rapidly. While science fiction often portrays AI as robots with human-like characteristics, AI can encompass anything from Google's search algorithms to IBM's Watson to autonomous weapons. Artificial intelligence today is properly known as narrow AI (or weak AI), in that it is designed to perform a narrow task (e.g. only facial recognition or only internet searches or only driving a car). However, the long-term goal of many researchers is to create general AI (AGI or strong AI). While narrow AI may outperform humans at whatever its specific task is, like playing chess or solving equations, AGI would outperform humans at nearly every cognitive task.

Why research AI safety? Would AI bring war when AI is continued to develop by developed countries? In the near term, the goal of keeping AI's impact on society beneficial motivates research in many areas, from economics and law to technical topics such as verification, validity, security and control. Whereas it may be little more than a minor nuisance if your laptop crashes or gets hacked, it becomes all the more important that an AI system does what you want it to do if it controls your car, your airplane, your pacemaker, your automated trading system or your power grid. Another short-term challenge is preventing a devastating arms race in lethal autonomous weapons.

In the long term, an important question is what will happen if the quest for strong AI succeeds and an AI system becomes better than humans at all cognitive tasks. As pointed out by I.J. Good in 1965, designing smarter AI systems is itself a cognitive task. Such a system could potentially undergo recursive self-improvement, triggering an intelligence explosion leaving human intellect far behind. By inventing revolutionary new technologies, such a superintelligence might help us eradicate war, disease, and poverty, and so the creation of strong AI might be the biggest event in human history. Some experts have expressed concern, though, that it might also be the last, unless we learn to align the goals of the AI with ours before it becomes superintelligent.

There are some who question whether strong AI will ever be achieved, and others who insist that the creation of superintelligent AI is guaranteed to be beneficial. At FLI we recognize both of these possibilities, but also recognize the potential for an artificial intelligence system to intentionally or unintentionally cause great harm. We believe research today will help us better prepare for and prevent such potentially negative consequences in the future, thus enjoying the benefits of AI while avoiding pitfalls.

How can AI be dangerous when developed countries continue to develop AI to become weapon to replace soldiers?

Most researchers agree that a superintelligent AI is unlikely to exhibit human emotions like love or hate, and that there is no reason to expect AI to become intentionally benevolent or malevolent. Instead, when considering how AI might become a risk, experts think two scenarios most likely:

The AI is programmed to do something devastating: Autonomous weapons are artificial intelligence systems that are programmed to kill. In the hands of the wrong person, these weapons could easily cause mass casualties. Moreover, an AI arms race could inadvertently lead to an AI war that also results in mass casualties. To avoid being thwarted by the enemy, these weapons would be designed to be extremely difficult to simply "turn off," so humans could plausibly lose control of such a situation. This risk is one that's present even with narrow AI, but grows as levels of AI intelligence and autonomy increase.

The AI is programmed to do something beneficial, but it develops a destructive method for achieving its goal: This can happen whenever we fail to fully align the AI's goals with ours, which is strikingly difficult. If you ask an obedient intelligent car to take you to the airport as fast as possible, it might get you there chased by helicopters and covered in vomit, doing not what you wanted but literally what you asked for. If a superintelligent system is tasked with a ambitious geoengineering project, it might wreak havoc with our ecosystem as a side effect, and view human attempts to stop it as a threat to be met. So, a super-intelligent AI will be extremely good at accomplishing its goals, and if those goals aren't aligned with ours, we have a problem. You're probably not an evil ant-hater who steps on ants out of malice, but if you're in charge of a hydroelectric green energy project and there's an anthill in the region to be flooded, too bad for the ants. A key goal of AI safety research is to never place humanity in the position of those ants.

Why the recent interest in AI safety ?

Stephen Hawking, Elon Musk, Steve Wozniak, Bill Gates, and many other big names in science and technology have recently expressed concern in the media and via open letters about the risks posed by AI, joined by many leading AI researchers. The idea that the quest for strong AI would ultimately succeed was long thought of as science fiction, centuries or more away. However, thanks to recent breakthroughs, many AI milestones, which experts viewed as decades away merely five years ago, have now been reached, making many experts take seriously the possibility of superintelligence in our lifetime. While some experts still guess that human-level AI is centuries away, most AI researches at the 2015 Puerto Rico Conference guessed that it would happen before 2060. Since it may take decades to complete the required safety research, it is prudent to start it now.

Because AI has the potential to become more intelligent than any human, we have no surprise way of predicting how it will behave. We can't use past technological developments as much of a basis because we've never created anything that has the ability to, wittingly or unwittingly, outsmart us. The best example of what we could face may be our own evolution. People now control the planet, not because we're the strongest, fastest or biggest, but because we're the smartest. If we're no longer the smartest, are we assured to remain in control?

A captivating conversation is taking place about the future of artificial intelligence and what it will/should mean for humanity. There are fascinating controversies where the world's leading experts disagree, such as: AI's future impact on the job market; if/when human-level AI will be developed; whether this will lead to an intelligence explosion; and whether this is something we should welcome or fear. But there are also many examples of boring pseudo-controversies caused by people misunderstanding and talking past each other. When one developed country continue to develop AI, can itself country's all factories workers will lose jobs, due to AI can replace them to do simple works in factories, or any public transport drivers, e.g. bus drivers, ferry , tram, train drivers, they will lose jobs, when AI (non manual driving drivers) can replace all public transport drivers. So, some occupations will lose if developed countries continue to develop or research AI to replace human to do some simple jobs, such as some cooking jobs can be done by AI. So, it is possible that future cookers won't be needed, because AI cooking skills may be better than them to cook any good taste chinese or western food in restaurants. If you drive down the road, you have a subjective experience of colors, sounds, etc. But does a self-driving car have a subjective experience? Does it feel like anything at all to be a self-driving car? Although this mystery of consciousness is interesting in its own right, it's irrelevant to AI risk. If you get struck by a driverless car, it makes no difference to you whether it subjectively feels conscious. In the same way, what will affect us humans is what superintelligent AI does, not how it subjectively feels.

In fact, AI may be make any brokers jobs in financial market. the main concern of the beneficial-AI movement isn't with robots but with intelligence itself: specifically, intelligence whose goals are misaligned with ours. To cause us trouble, such misaligned superhuman intelligence needs no robotic body, merely an internet connection – this may enable outsmarting financial markets, out-inventing human researchers, out-manipulating human leaders, and developing weapons we cannot even understand. Even if building robots were physically impossible, a super-intelligent and super-wealthy AI could easily pay or manipulate many humans to unwittingly do its bidding. So, future brokers will be replaced by AI, when AI can be made to own financial brokers' analytical mind to make more accurate whether the share price will rise up or fall down to compare human financial brokers' analytical mind. The robot misconception is related to the myth that machines can't control humans. Intelligence enables control: humans control tigers not because we are stronger, but because we are smarter. This means that if we cede our position as smartest on our planet, it's possible that we might also cede control.

Not wasting time on the above-mentioned misconceptions lets us focus on true and interesting controversies where even the experts disagree. What sort of future do you want? Should we develop lethal autonomous weapons? What would you like to happen with job automation? What career advice would you give today's kids? Do you prefer new jobs replacing the old ones, or a jobless society where everyone enjoys a life of leisure and machine-produced wealth? Further down the road, would you like us to create superintelligent life and spread it through our cosmos? Will we control intelligent machines or will they control us? Will intelligent machines replace us, coexist with us, or merge with us? What will it mean to be human in the age of artificial intelligence?

Why do developed countries people need AI ?

Why do we assume that AI will require more and more physical space and more power when human intelligence continuously manages to miniaturize and reduce power consumption of its devices. How low the power needs and how small will the machines be by the time quantum computing becomes reality? Why do we assume that AI will exist as independent machines? If so, and the AI is able to improve its Intelligence by reprogramming itself, will machines driven by slower processors feel threatened, not by mere stupid humans, but by machines with faster processors? What would drive machines to reproduce themselves when there is no biological incentive, pressure or need to do so?

Who says superior AI will need or want to have a physical existence when an immaterial AI could evolve and preserve itself better from external dangers. What will happen if AI developed by competing ideologies, liberalism vs communism, reach maturity at the same time, will they fight for hegemony by trying to destroy each other physically and/or virtually. If AI is programmed to believe in God, and competing AI emerges programmed by muslims, christians or jews, how are the different AI's going to make sense of the different religious beliefs, are we going to have AI religious wars? What if the "powers that be" greatest fear is the emergence of a super AI that police's and rationalizes the distribution of wealth and food. A friendly super AI that is programmed to help humanity by, enforcing the declaration of Human Rights (the US is the only industrialized country that to this day has not signed this declaration) ending corruption and racism and protecting the environment.Most benefits of civilization stem from intelligence, so how can we enhance these benefits with artificial intelligence without being replaced on the job market and perhaps altogether?

Key to the process of machine learning are neural networks. These are brain-inspired networks of interconnected layers of algorithms, called neurons, that feed data into each other, and which can be trained to carry out specific tasks by modifying the importance attributed to input data as it passes between the layers. During training of these neural networks, the weights attached to different inputs will continue to be varied until the output from the neural network is very close to what is desired, at which point the network will have 'learned' how to carry out a particular task. A subset of machine learning is deep learning, where neural networks are expanded into sprawling networks with a huge number of layers that are trained using massive amounts of data. It is these deep neural networks that have fuelled the current leap forward in the ability of computers to carry out task like speech recognition and computer vision.

In conclusion, when developed countries continue to develop AI, it may bring positive advantages to bring raising productivies, or efficiencies, but it may also raise unemployment ratio to any low skill or low knowledge jobs in ther societies. However, human future society will need to change to be better to raise our living standard. But AI is one kind the best choice tool to achieve this aim in our future, so I agree developed countries continue to develop or research AI to be the super -human machine.

Reference

A. Castano et. al. " Automatic detection of dust devils and clouds at Mars" Machine vision and applications, Oct. 2008, vol. 19, no 5-6, pp. 467-482.

Accenture, " Why artificial intelligence is the future of growth"(2017) <http://www.accenture.com/us-en/insight-a   rtificial-intelligence-future-growth>.

D. Schedidt , Unmanned Air Vehicle Command And Control, Handbook Of Unmanned Air Vehicles, Springer-Verlag, 2014. Facebook (AI) Research Available at https://research.facebook.com/ai, research at google, machine intelligence available at http://research.google.com/pubs/machineintellige nce.html; micro soft research-machine learning and artificial intelligence available at http://research.microsoft.com/en-us/research-   areas/machine-learning-ai.aspx.

International Federation Of Robotics, 2016. IFR press release world robotics report. IFR, org . 29 Sept. Accessed Feb. 01, 2017. http://www.ifr.org/news/ifr-press-release/world-robitics   report -2016-8321.

K, Fedra , "GIS and environmental modelling" in environmental modelling with GIS, edited by M.F. Goodchild.B.O. Parks and L.T. Steyaert, Oxford University press, pp. 35-50, 1994.

Keynes, J.M. (1933). Economic possibilities for our grandchildren (1930). Essays in persuasion, pp.358-73.

Mckinsey & Company (2013, May). Disruptive technologies: Advices that will transform life, business and the global economy , USA.

Ministry of economy, trade and industry, Japan, 2015, Japan's robot strategy. Ministry of economy, trade and industry.

Ray Kurzweil , The age of spiritual machines (1999) is cited numerously through this chapter: Kurzweilai.net http://www.kurzweilai.net

Rich, Elaine & Knight, Kevin, Artificial Intelligence Second Edition, 1991, New York; Mc-Graw-Hill.

Skills shortages on developing country market

Future developing countries need to develop their economy, so they need to employ many employees who own technical skills and /or soft skills. What kind of technical skills and/or soft skills , the developing countries' employees who will need to order to raise competition in local job market ? I shall indicate the developing country China example. China is one developing country, employers will need different kinds of skillful labors to assist them to develop their businesses. However, China employers will face skillful labour shortage challenge. Although Chinese young age population is high, but many of them do not to be encouraged to learn enough skillful knowledge to fill future new skillful positions. So, the fast speed of training will be important to influence China supply and demand labour market to be more accurately as well as future China's the quality of labour demand number will be influenced to be raised after they have enough training to learn new skills.

How to solve future China skillful shortage of labour? Firstly, nowadays, China employers need to teach their employees to learn how to use and how to operate robotic skills in China's factories. AI robotic has been early developing, so they need to prepare to learn robotic management and operating technical and soft skills in order to satisfy future China factory automation industry development.

China is one world's factory for low-end products to high quality information products, high end technology and services. So, China will need many high skilled workers to assist manufacturers to manufacture many different kinds of products to export or local sale. Moreover, robotic manufacturing skillful workers will also need because robotic will be accepted to assist manual workers to work in China's any factories. This has led to greater demand for labour with upgraded skills and competence. So, it seems that China's workers need to learn any high technological manufacturing knowledge, e.g. learning how to co-operate with robotics to raise productive efficiencies, which will be future many China's manufacturers' skills need intention.

So, when any one of China manufacturer invests robotics to work in its factory . Then, the China manufacturer's labours ought need to know how to co-operate with the robotics to raise productivities and efficiencies. Moreover, these China service industries, e.g. IT, software, accounting, finance, marketing and customer service management, e.g. waiter, property security, shopping center customer service etc. service occupations. In the future, robotics can also used to participate any one of these service industries' part of tasks in order to raise service performance. So, any one of these service industries' employees need to learn how to operate with robotics in order to achieve the most excellent service performances to satisfy consumers' needs. So, China service industries labours ought need to learn how to co-operate or manage service natural robotics to work together more efficiently because future China manufacturers will prefer to employ the labours who know how to co-operate and manage and control any service natural robotics more easily and efficiently in order to achieve the most excellent service performance to satisfy customers needs.

Hence, it seems that China manufacturing and service workers need to spend time and effort to learn how to co-operate with manufacturing natural robotics to manufacture any products in factories efficiently or deliver any cargos in warehouses more efficiently or serve customers to let them to feel excellent service performance in restaurants or shopping centers or properties or offices reception counters. Then, when their China employers apply robotics to participate to work in factories, restaurants, shopping centers, cinemas, offices or properties reception etc. different working places . These low skillful labours will be dismissed easily, due to robotics can replace them to manufacture any products or provide services to satisfy clients' needs in order to let them to fell robotics' performances are more excellent to compare human service labours or their productive efficiencies are more effort to compare workers. So, future China workers need to learn how to cooperate or manage or contol with robotics to work more efficiently, if they do not expect to be dismissed easily.

In the future several occupations have been identified as the most frequent movers between all labour market states. The elementary occupations include: waiters, bar staffs, clearners, catering assistants, construction and security service workers, care workers, sales assistants and general clerks etc. So, the low educational level workers can learn these soft wkills to raise whose professional workering level to prepare to do these above positions in global elementary occupation job market.

The changes of employer were most frequent for IT programmers, doctors, electricians, carpenters, skilled workers in global labour market. These skilled occupations will have manpower shortage supply challenge, due to either people feel the educatonal level is under low. So, there has no many people have interest to know these knowledg to prepare their elementary careers. So, these kinds of low skilled occupations will have not enough human power supply to global labour job market also, the high skilled or educational job support.

Moreover, the high skilled occupations also encounter labour shortage issue. The skills in short supply related to experienced canadidates e.g. five years or more. For example, pharmaceutical , biogharma and food innovation industries. The occupational shortage roles include: Chemists, analytical scientists, product formulation, analytical development for roles in biopharma, quality control analyst includes pharmaco-vigilance, i.e. drug safety roles. The demand for engineering industry aspect which will aos increase the labour shortage includes process and design (research and development, quality control, automation, lean processes) are skillful labours need to help employers to achieve these intentions. They may include raising competitiveness, boosting productivity and skills availability. So, if future these above any one of occupation labours can not achieve these benefits to satisfy their employers' needs. Then, his/her average weekly or hourly wages will be reduced. It means that the unskilled labour under skilled labour wage can not increased more easily, even they own many year working experiences in any one of above these occupations.

If the employer feels the labour is unskilled or below skilled level for any one of these occupations in these any one industry aspect, e.g. wholesale and retal , human health, education, accomodaton and food , construction, professional activities, financial service , public administration, and defence, transportation etc. occupations. Then, these industries' unskilled or below skilled level workers' salaries will be lower level to compare the higher skilled workers who work in any one of these industries.

The reason why future employers need to employ skilled labours. One explanation for slow recovery in demand in negative impact on investment is a prolonged period of high unemployment. This is led to job weekers left labour market or became unemployable due. So, future low skillful level will be one important factor to cause unemployment in society as well as nowadays labours ought need consider whether their skills are needed to improve in order to avoid future competition in job market.

2.1 Why do future labours need to learn worldwide readiness skills

Future employers need employees own worldwide readiness skills, such as reading , writing and arithmatic. Why do employees need worldwide readiness skills? In the future, high economic growth countries need high wage positions, high opportunity jobs which need a large number of skills required of job candidates of these positions " job readinss" and not " job training" , which support developments of these importance and widely desired skills won't only support the success to high-opportunity positions, but also be developed for future success in the competitive global economy. Because real-time business intelligence is needed for the talent marketplace to employ talent employees. So, it explains that it will have many future employers hope to employ owning readiness skillful employees to help them to develop their businesse intelligently. Hence, present employees ought need to hard to train readiness skills to prepare whose future employers' job requirements in the future competitive global job market.

2.2 Why these occupations need readiness skills

In the future these occupations will need to raise readiness skills. For example, mathematical science, teachers (post-secondary), management analysts, computer and information systems, managers, first-line supervisors of construction traders, solar photovoltaic installers. All of these occupations , employers need staffs to own good readiness analytic ability to help them to do more accurate real-time business intelligent decisions. The representative occupations include oral and written communication skills, project management, teamwork, marketing and creativity . Moreover, they need to own specific technology skill, deep science and math or even most business skills as well as these skills are "soft" skills more than hard skills. These kinds of occupation employees need own cooperative effort, creativity, problem solving, detail orientation and integrity personal characteristics, which are relevant across all knowledge and domains.

Therefore, in the future, science, technology,engineering and mathematics relevant occupations need to own more read and analytic skills more than other kinds of occupations. Because these organizations need those professionals on knowledge acquisition, literacy analysis, synthesis and critical thinking skills that will impact their organizations to bring more critical thinking beneficial team culture. These occupational top skills will include oral and written communication skills, project management skill, team oriented skill, marketing and creativity skills, problem solving skill, detail oriented skill, self-motivated skills, management and analytical skills, coaching skill, business process modeling skills, work independent skill, strong leadership skills, management experience and business requirements gathering. All of these skills which will be future employers who need to employ these kinds employees who own these skills in preference. Also, all of these skills concentrate on soft skills more than hard skills. It seems that when above occupational applicants who own any one of these skills, even more than one skills. Then, he/she will have more chance to be selected to employ. Also occupation specific skills requirements are more needed to compare cross-functional skills for above of any one occupation. Because the high concentration of cross-functional skills require " job readiness" and not " job training" for success, e.g. communication, integration and presentation skills, entrepreneurialism and related skills, microsoft office software skills.

Of particular interest is communication, integration and presentation skills. These skills include ability to seek, evaluate and examine information and data create a reasoned position, present findings and make a case for or advocate for position. So, these skills are very important and they can help future applicants who expect to win any kinds of these positions easily. However, the hard skills can help these applicants to be more successful to win any kinds of these positions when they own these hard skills, e.g. microsoft office, powerpoint, excel , word, microsoft project etc. softwares. In conclusion, the global economy is dynamic and many of the skills required for positons in the future will need good technologies and work practices to be developed. The number of skills required t be successful in the jobs forecast to be most in demand in the future is growing. So, it explains that why future any one of these occupations which will need soft skills more than hard skills, due to organizations like to employ the employees who own managerial and analytical effort more than hard skills productive effort to assist their organizations to develop more easily.

2.3 Data -analysis skill needs

In the future, most organizations will have a number of jobs that include data analysis. Economists and labor market forecasters predict occupations need data analytical skill will need much. In addition, fast technological development means th types of technologies and applications workers in this field will need to be familiar with data analytical skill rapidly. It seems that data analytical jobs will have new job opportunity to employees with in-demand skills in future global labor market.

Why and how do employers demand for data analysis skills? Data analysis skills mean the ability to gather, analyze and draw practical conclusions from data as well as communicate data findings to others. The occupations include: data analyst, data scientist, statistician, market research analyst, financial analyst,research manager. In business career, many employers expect to employ statisticans, operations research analysts, market research analysts and marketing specialists to assist their organizations to gather useful data from market in order to analyze and draw practical conclusions and finding the best solutions or methods to win their competitors.

Therefore, these data analysis jobs will have much need. Large size organizations with 500 or more employees were more likely than small or medium size organizations with 25 to 499 employees to plan hired data analysis positons in the future. For example, human source department will use big data to help make strategic decisions. How HR uses big data . HR will use big data for sourcing, recruitment, or selection, identifying causes of turnover and/or employee retention strategies or trends, managing talent and performance. Why organizations do not use big data. It is possible that they lack of knowledge expertise, the majority of organizations will have data analysis positions within accounting and finance department, human resources department, business and administration department, information technology department, marketing, advertising and sales department, supply chain and operations department, research and development department, customer service department and other departments. So, future data analysis skill will need to used in different organizational departments.

However, publicly and privately owned for-profit organizations were more likely than government organizations to have data analysis positions in the marketing, advertising and sales function. Also, data analysis skills are required to different levels in any organizations , such as entry level, non-management / individual contributor level, mid-level management level, senior management or executive level. The analyst, research analyst, market research analyst, scientist-based titles include: data scientists , research scientist, scientist, other descriptive titles include researcher, statistician, mathematician and other . So, data analysis positions will have many different skills to be selected to any one data analysis professional. For example, the data analysis professional can select either to learn the ability to interpret and communicate data analysis results skill or to learn how gathering or analyzing data skill. So, data analysis skill is not only one skill, it is more than one skill to let any one employee to select to learn.

Why do organizations need data analysis professionals? On workforce planning aspect, organizations expect to let strategic direction and content of workforce needed for future business objectives easier, analyzing workforce: supply analysis, demand analysis and gap analysis more easy, developing action plan : recruiting and training plans to deal with gaps more easier, implementing action plan, monitoring, evaluating and revising plan more easier. So, organizations expect the data analysis professional can help them to solve these challenges, such as using of advanced technology solutons to integrate disparate planning sources; data availability and format; accessing to and understanding of the organization's data and analytics, developing business case to gain support from senior management and collaboration among HR staff, managers and executive easier. Future industries need data analysis professionals may include manufacturing health care and social assistance, scientific and technical service, finance and insurance, educational services , government agencies, retail trade, transportation and warehousing, construction, utilities, accommodation, and food services, waste management and remediation services, entertainment, and creation, real estate and rental and leasing , repair and maintenance, agriculture, forestry, fishing and hunting, personal and laundry services etc.

In conclusion, data analysis job need explains why future readiness and data analytical skills will be popular needed in global labour market , due to these both skills are labour shortage and employers will need employees own big data readiness and data analytical both skills in order to win whose competitors more easier.

2.4 What are regional dynamic skills
of global labour market demand when robotic participate labor market

Businessmen expect to improve better economic environment, they will prefer to recruit the most sought after skills of intelligent employees to bring positive beneficial impact to organizations. However, technology and digization has had a significant influence on workers. Future globalization will trend digital economic development. Hence, it will influence workers‘ skills to be changed also. In fact, not all changes are positive because some workers will possible lose jobs, either due to new technology replaces their jobs or they lack enough effort to improve their skills in global digital economic labour market environment.

It brings this question: What are regional dynamic skills need with digital business environment is growing. In fact, organizations will continue to deal with skills shortages, labour markets across the global are continually changing. so, more employers and workers will need to adopt innovate working pattern, e.g. on call jobs, freelance jobs will grow popularly. The greater flexibility afforded to employ regard.

Finally, digitalisation includes artificial intelligence, big data , online platforms. All these new technology will intluence tuture employees how to worker. For example, they can apply online platform to work at home conveniently. So, they do not need to go to offices. They can finish their jobs and send to their employers by email easily. This kinds of job pattern can raise efficiencies and employers do not need go to offices often.

An important implication of innovating working which needs the employees who own digital skills in order to serve organizations more efficiently. So, employers are increasingly able to access demographics that were hitherto less active in labour markets. For example, future more women are joining the labour market because part time and self employment opportunities make it easier. This kinds of job pattern can raise efficiencies and employees do not need go to offices often.

An important implication of innovating working which needs the employees who own digital skills in order to serve organizations more efficiently. So, employers are increasingly able to access demographic that were hitherto less active in labour markets. For example, future more women are joining the labour market because part time and self employment opportunities make it easier to manage family with work life. So, digital skilling needs will cause many women lose jobs in possible. If the women lack digital job skills. Because high digital skill occupations need, like those requiring research, medical treatment and architectural design occupational digital skills are more common in the services sector, more women who own digital skill who can compete to win.

High digital skill occupations more easier than men because employers usually select female to do high skill occupations easier than make. However, if those professional service female employees can not learn how to apply digital skills to do these researchs medical treatmentm architectural design professional service jobs. Then, it is also different for these professional service femal employees to raise competition in global labour professional service market. So, these professional service female employees need to learn how to apply digital to do themselves jobs in future global professional service labour market. Otherwise, if the male professional service employees can attempt to learn how to apply digital skill to do themselves jobs in order to improve efficiencies and service performance to satisfy patients, such as medical service needs, school search service needs, construction firms' building needs. Then, the owning high digital technology skillful female employees will be more easier to find the professional service jobs which need digital skill more easier than the lacking digital skill female service professionals in future global digital service professional labour market.

On the other robotic communication skill need aspect, future employers expect workers to know how to communicate with robots to work efficiently in any working environment if the employers need robotc to serve their organizations. For example, communication between the robots on factory floors, and between people and robots could allow robots to start and stopr processes based on real-time conditions around them and alert people when there is a problem, so robots could increase their own efficiency if the workers could monitor themselves and determine when they needed maintenance; efficiency would also be improved if machines and robots could make production decisions on their own by. For example, ordering new suppliers when existing inputs into a production process run low. The increase in productivity of industrial robots will likely reduce the number of manual jobs on the shop floor.

At the same time, the increased output made possible by such robots will mean that manufacturers need more people in accounting, finance, sales, advertising and other roles. The increase in putput may also drive increased employment on manufacturers' supply chains. Hence, future employers expect to employ the workers who can know how to communicate with robots to work efficiently in order to raise productivity in any working environment. It means that it the worker can know how to control and communicate with the robots to work together in the team. Then, his/ her communication and controlling robotic skill will help the organization's team to work efficiently and raise productivity in order to reduce time waste and human waste and resource waste considerately. So, future shortage of communication and controlling robotic skillful workers number will increase. It has much beneficial to workers who choose to attempt to learn how to communicate and control robots to work together in any working environment team efficiently. Because future employers will like to use robots to assist manual workers to attempt to raise productive efficiency in any working environment. So, the need of employees who know how to cooperate or communicate with robots whose talent skills will be useful to any future employers.

Future global business leaders will need human machine cooperation skill. This technological skill includes artificial intelligence (AI and internet of things (IOT), will reshape our working change. These machines will participate to our daily working environment. For instance, many business leaders agree that automated systems will free-up their time as well as they also believe they'll have more job satisfaction by offloading the tasks that they don't want to do to intelligent machines.

Therefore, future leaders will expect humans and machines can work as integrated teams within their organization in order to their workforce and machines are already successfully working this way. So, they need to expect future employees can know or learn how to work with automated systems more easily, because many jobs will be participated by automated systems, e..g simple accounting tasks, legal administration tasks etc. clerical tasks. They will be participated with (AI) technology, it learns how to cooperate with (AI) technology to finish simple clerical tasks efficiently.

Future workers will need have automated system operational skills: They include that how to operate automated systems to free -up workers' time. Workers will need to learn how to operate automated system to better with healthcare tracking devices workers will need to learn how to operate automated systems to absorb and manage information in completely different ways. Workers will need to learn how to operate automated systems of smart machines to work as admin. in any working environments. Workers need be needed to learn how to operate (AI) automated machines to make more accurate clerical tasks or efficiencies. So, the automated system (robotic) operational skillful workers' demand and number will increase.

In the future, employers need automated machine manufacturing and service with workers cooperation reasons include that clear protocols, will need to be established if autonomous machines fail. So, they need their workers to learn how to control and manage and communicate with autonomous machines skillfully. They believe move they depend upon technology, the more they'll have to lose in the event of a cyber attack. So, skillful workers are real required to let them to know how to cooperate with autonomous machines more efficiently and easily. Computers will need to be able to decipher between good and bad commands, so future employers have much chance to need the owning automated machines operating workers to assist any robots to make more accurate good or bad decision when robots and workers have need to make immediate judgement in their any related job responsibilities aspect.

Therefore, future owning automated machines operating workers' skillful level will be high. It bases on automated machine manufacturing environment trend factor. Finally, future technology will connect the right employee to the high task at the right time. It implies that when future global employers began to accept to apply robots to help them to raise any productivities efficiently. It will influence many manufacturing positions which need to employ any proficient skillful workers who own automated machines operational skills to know how to communicate or manage or control , even supervise any robots to work in teams in any organizational manufacturing environment efficiently.

In the future, employers also expect employees to own sufficient digital vision and strategic skills, manifest among other things. They can know how to apply data to demonstrate any senior support and sponsorship digital technological skill. They expect to reduce a skill gap and avoid a lack of employee buying and a workforce culture to change in their digital technological manufacturing organizations. Future employers also believe outdated technology that can't work fast enough, data overload, privacy and security concerns. So, it explains why it is possible that future employers also need digital working environment and automated robots machines to attempt to achieve raising productive efficient aim.

Moreover, it also explains why digital transformation need will be raised. The reasons include: They feel digital technology can gain employees' buying in , making customer experience a boardroom concern, achieving fair compensation , training and goals and strategy achievement more easily, tasking senior leaders with digital working environment change putting policies and technology to support a fully remote, flexible workforce , empowering lines of team work more efficient, teaching all employees how to code/understanding how to adopt to work with automatic machines or rots in any team efficiently. So, automate machine can raise efficiency in manufacturing society.

In conclusion, in the future business society, employees need to be stronger human machine partnerships. So , future manufacturing or service industries will have digital technology and automated machine robotic technology to assist workers to work in any working environment efficiently. They expect digital technology and automated machine robotic technology anticipation to workers' daily jobs in order to bring positive impacting to the customer experience from business owners to decision makers in marketing, customer service, research and developmnt and finance etc. They also expect technological productivity can bring positive relationship between technology and workers emerging technologies' impact on business and the way workers and automated machine work together.

In the future whether in general organizations need what kinds of employees' skills, they expect employee individual own. It is one interesting question. The common skills that employees need to own in order to any duties to any organizational departments efficiently, e.g. human resource, marketing, administrative, logistic etc. different departments. For hospital, school, business, professional occupations etc. different organizations. Whether future school ought implement one system educational method to teach different common skills to students in order to let them to leave schools to jobs more easier.

Future employers need to create new technologies including automation and algorithms, in order to create new high quality jobs and improve the job quality and productivity of the existing work of human employees in any organizations, e.g. accounting department will need intelligence (AI) to assist account clerks to do simple repeating accounting job tasks in order to share their work load and raise performance efficiency or legal organizations will need (AI) to assist law clerks to do simple repeating legal draft or legal document revising job tasks . All future general clerical jobs will apply (AI) technological tools to assist human to job, it will produce a comprehensive platform for managing workforce change.

Hence, human manual(employees) need to learn how to adopt (AI) job participation to assist them to do different kinds of simple clerical jobs in any organizational administrative departments . They , clerical employees or white color workers need to learn how manage or dominate (AI) tool to improve job performance to be better. However, (AI) administrative workforce change, it is not only one kind of job automation change role in any physical offices. It influences future administrative clerks need change a more flexible manner, utilizing remote staffing beyond physical offices and decentralization of operations organizational workforce change.

Instead of (AI) participation to administrative job aspect, (AI) will also participate to manufacturing industry environment aspect, a new human-machine manufacturing workforce change will exist to any factories, warehouses working environment. Scientists predict that in present an average of 71% of total task hours across the industries are performed by humans, compared a 29% by machines. In this average is expected to have shifted to 58% task hours performed by humans and 42% by machines. In fact, nowadays, in terms of total working hours, no work task was yet estimated to be predominantly performed by a machine or an algorithm (AI). But, this picture is predicted to have somewhat changed with machines and algorithms (AI) on average increasing their contribution to specific tasks by 57% . For example, in the future, 62% of organization's information and data processing and information search and transmission tasks will be performed by machines compared to 46% today.

Therefore, these high technological skillful job change will bring negative influence to some demotive-skillful or low skillful labors to be dismissed, if they can not upgrade or raise or re-skillful their skill level to improve their analytical thinking , technology design and programming skills to cooperate with (AI) tools to work efficiently together in any organizational manufacturing or office work environment. Because it will have many employers apply (AI) automation tools to participate with blue -color or white -color workers' tasks in order to raise efficiencies or improve performance in any working environment. So, it is right time to young or mid age employees need to upskill and/or reskill their rihgt type of skills to prepare future technology rich work environment changing needs.

Future technological advances will permit an increasing number of tasks traditionally performed by humans to become automated. It seems that , such automation focused primarily on routine tasks, e.g. clerical work, bookkeeping, basic paralegal work and reporting etc. However, with the advent of big data, artificial intelligence (AI), the internet of things and ever-increasing computing power , i.e. the digital revolutions, non-routine tasks are also increasingly likely to become automated. For example, the recent development in robotics and 3D printing allow firms in advanced economies to locate production closer to domestic markets in fully automated factories. As a result, the future strongest incentive to automate because of their relatively higher labour costs will be reduced, when production automated will bring the negative influence to dismiss some foolish or low productive or low skill workers , the owning high automated productive skillful workers will replace the low productive skillful workers in any factories' manufacturing environments. So, technological progress participates to raise quantity of jobs will cause result in significant job losses to low skillful workers. Because future employers will need many high automated productive employees to help them to cooperate with (AI) automated machine to work together efficiently. For example, many proportion of occupations at high risk is greatest in Germany and lowest in Korea, these countries organizations will accept to spend technology investments and education of workers to prepare future automatability manufacturing development successfully.

However, future automat ability manufacturing development will bring technological unemployment in possible, due to workers need to adjust to the challenge of automation by switching tasks. Thus, preventing technological unemployment, also technological change does not just destroy jobs, but also generates new roles through its effect on productivity and the demand for new technologies. For example, it has been estimated that, for each high tech-job created in the industries , such as computing equipment or electrical machinery, some 4.9 % additional jobs are created for lawyers, taxi, drivers and waiter in the local economy (Moretti, 2011).

Therefore, automated will also influence service industries' job nature change, e.g. taxi drivers need to apply (AI) automated machines to assist them to drive their taxis. When the passenger tells the taxi driver where he/she wants to go. Then, the (AI automated machine will follow the GPS road direction map to be indicated how to drive the taxi to go to the destination automatically . So, future taxi driver is one assistance role to assist the (AI) automated driving tool to dominate the (AI) tool to drive the taxi to catch the passenger to arrive the destination safety in the short time in possible. For another example, future restaurant waiters will need (AI) automated machines's assistance to help them to deliver or dispatch any foods and soft drinks to send to the identified eater's table carefully in accurate and efficient service performance way from the kitchen, in especially in the busy time and many people are sitting in the large size restaurant environment. So, future, waiter roles will be the leader , they need to manage or control or supervise the (AI) robotics how to make decisions to arrange to dispatch which foods or soft drinks to the different tables in preference immediately. Also, future law clerks need to supervise or manage the law robotics how to help them to make decisions to do revision or draft or filing legal tasks in preference in order to avoid any typing words are mistaken to type on computers or revised draft in wrong way to assist manual legal clerks' mistaken words are appearance on any legal documents. So, the law clerk future role will be the trainer role , he/ she needs to teacher the robots how to check any words, e.g. grammars to correct them to be right grammars, or giving the accurate revision legal documents' instruction to let the legal robots to know how to revise each legal draft to prove whether which part of the legal draft will have wrong to be needed to revise.

In conclusion, future many manual workers' service or manufacturing job natures will become automated assistance to robotics. So, employees need to upgrade their skills in order to adopt new technological work nature change.

Reference

Moretti, E. (2011) local labor market in O, Ashentelter and D. Card (eds.) handbook of labor economics, Elsevier, North Halland.

Becoming robotic talent human how influences social changes

Nowadays we tend to think about social and digital technology more from a personal or consumer perspective than their business or professional applications, but as the Digital Era continues to progress, many of technology's most profound impacts are likely to be in the world of work. In addition to changes in product and business development, knowledge management, data analysis, and other operational processes, transforming talent management will be a key priority for organizations striving to be employers of choice.

● Digital technology encourages to create talent human

Why does digitized technology encourage to create talent human or excite human to learn new things ? The human capital implications of social and digital technologies impact virtually everyone, regardless of the type of organization they work for, their profession, their functional area, or their career stage. That means that the talent management functions in all organizations, as well as the professionals who staff and lead them, have a critical role to play in ensuring the efficient and effective transition and transformation from Industrial Era models and processes to their Digital Era upgrades.

It's no surprise that talent management has already become more "high tech." Many employment related activities have been digitized, and there has been a corresponding increase in employee self-service. It's important to remember, however, that digitization is not the same thing as digital engagement, and that the rise of "high tech" solutions doesn't necessitate the loss of a "high touch" approach to managing an organization's human assets. Transforming talent management requires digitization, to be sure, but it also involves leveraging social and digital technologies in ways that promote and enhance communication, collaboration, and engagement - not just between an employee and the organization, but between and among employees themselves.

Talent Acquisition

The logical place to start when talking about the impact of social and digital technologies on talent management is talent acquisition, where the greatest advances have been made. Anyone who has searched and applied for jobs in the past 10 years is very familiar with how technology has transformed the application process, which in most organizations (and virtually all large ones) is now almost completely digitized and automated. However, there are other ways in which social and digital technologies are impacting talent acquisition that may not be as well-known or commonly understood. Social media sites in particular (such as Facebook, YouTube, and Pinterest) are a great way to promote an employer's brand and offer realistic previews of work life, people and culture in organizations. Online games and simulations can also be used to get a sense of what working for an organization would be like, and give organizations themselves an opportunity to determine if a prospective candidate would be a good cultural fit and potentially successful.

On organizational working environment aspect, some employers are recognizing the value of digital alumni networks or communities to maintain strong relationships with former employees. One of the primary motivations for doing this is that the employees may return one day and/or make referrals to or from their personal and professional networks. Similarly, talent networks enable organizations to establish and maintain relationships with professionals in key areas like IT and engineering, even when there isn't a current opportunity to have those folks be a part of the organization. Moreover, social media can bring positive influence to impact organizations to encourage employees to attempt or feel needs to learning new things for their tasks needs. Due to social media has actually transformed every stage of the recruiting process in significant ways - so much so that the traditional recruiting funnel can be recast in "social" terms. At the top of the funnel are activities like social advertising (i.e., placing job ads on social networks like Facebook), social sourcing (i.e., searching for candidates who meet certain criteria on networks like LinkedIn), and social referrals (i.e., having current employees share position openings with their online personal and professional networks). And at the bottom of the funnel is social screening (i.e., reviewing a candidate's public activity in social networks to identify potential hiring risks).

● Learning management is needed to feel needs to any organizations

Learning management is probably the another most advanced area when it comes to adopting and adapting to new technologies. As with recruiting and other processes, the initial advances are in the area of digitization, with social software applications evolving next. One of the obvious digital impacts is the increased use of el-earning and online learning platforms with self-paced study. There are also countless instructional videos on the web, both free and fee-based, that address a virtually unlimited range of topics. And we can't forget MOOCs - massive, open, online courses - which have proliferated in the past couple of years. Finally, many organizations have also started to leverage tablets and other mobile devices for learning, as well as using simulations and games to help employees develop specific skills. In addition to offering training through a variety of multimedia channels, organizations are increasingly using a range of digital tools for assessing employees' skills. They're also allowing employees to play an enhanced role in identifying their key skill sets and training needs, and can even have them create their own learning and development plans.

Allowing employees to take a more active role in their own learning and skills management enables organizations to develop and maintain a more complete and accurate knowledge and skills database, which in turn enables them to maximize the value of the workforce in which they've already invested.

1. Formal learning management systems and platforms are also beginning to incorporate social technologies in a variety of ways. Promoting connections and interactions among participants, as well as with the instructor, can enhance the learning experience both during and after a course. Creating course-based cohorts that allow people to continue to interact with each other via a digital community - even when their shared learning experience is face-to-face - can promote both knowledge transfer and retention, in addition to increasing commitment and engagement through interpersonal connections.

2. Informal learning - which is now also referred to as social learning - is greatly enhanced by social technologies as well. In fact, this is probably the greatest opportunity and area of growth for organizations of all types and sizes. Through private social networks, intranets and other internal platforms that have incorporated social technology elements, organizations are better able to facilitate employee learning as they perform their job duties and complete work activities. Along with the networks themselves, features like advanced search, identified subject matter experts, digital communities of practice, wikis and more enable employees to access and learn from colleagues who are not just next door or down the hall, but even in another city, state or country!

As organizations move forward with leveraging technology to enhance learning initiatives, it will become increasingly important for them to address issues related to digital literacy and digital competencies. For the past several decades we've generally taken what I refer to as an LIY, or Learn It Yourself, approach to digital knowledge and skills. Although organizations may invest in teaching someone how to use a specific application related to their job, they make virtually no investment in helping individuals learn how to use general digital tools like Microsoft Office and even email. Left to their own devices, most people - and I include myself in this group- are much less efficient and effective at using these tools than they could or should be. As our tools get even more sophisticated, we need the foundational knowledge and skills to be able to use them well - and this foundation should probably be provided via more formal training. In other words, many people need to be "taught how to learn" in the Digital Era. If organizations aren't going to provide the formal training workers need to do that, it's probably in an individual's best interests to pursue those kinds of development opportunities on their own.

● What are social influences on human behavior when talent human number increases?

Social Influences on Human Behavior Because human beings are social and learn from observation rather than depending entirely on instinct, almost all aspects of human psychology and behavior are socially influenced. Languages, modes of dress, gender roles and avoided taboos are all agreed upon at a group level and form the basis of culture. What are the characteristics of social change? Small-scale and short-term changes are characteristic of human societies, because customs and norms change, new techniques and technologies are invented, environmental changes spur new adaptations, and conflicts result in redistributions of power. This universal human potential for social change has a biological basis.

This universal human potential for social change has a biological basis. It is rooted in the flexibility and adaptability of the human species—the near absence of biologically fixed action patterns (instincts) on the one hand and the enormous capacity for learning, symbolizing, and creating on the other hand. Because human beings are social and learn from observation rather than depending entirely on instinct, almost all aspects of human psychology and behaviours are socially influenced. Languages, modes of dress, gender roles and avoided taboos are all agreed upon at a group level and form the basis of culture.

On conclusion, when one country can create many talent human, e.g. students, workers. Then they can bring more positive attribution to help or assist themselves country to develop rapidly. Consequently, the country's economic growth speed will be rapid. So, I believe that it has close relationship between economy growth and talent human number to any countries in nowadays societies.

Robotics can improve internet technology development to raise economic growth factors

Why do we improve to improve internet technology? What long term social benefits will benefits if scientists can improve internet speed and research any information function ? I shall research these questions to give suggestion as below:

Why does internet improvement make life better? Internet of Things Benefits In short, the scale of change that IoT technology offers can be scary. At the same time, the benefits of a well-executed IoT strategy can be more need for an organization: Safety, Comfort, Efficiency. Also, the Internet offers teens the ability to make friends with peers with whom they would not otherwise connect. With pop culture deteriorating into many distinct subcultures, teens' interests are more variable than they have ever been. With internet communication, employees can effortlessly communicate with one another at anytime from anywhere in the world. This allows employees situated in different parts of the world to give their opinion and voice their concerns. Through internet access, individuals in developing countries are able to gain access to more of the modern economy. With internet connectivity, those living in remote areas can now easily take out microloans, participate in e-banking and more. A large share of respondents predict enormous potential for improved quality of life over the next 50 years for most individuals thanks to internet connectivity, although many said the benefits of a wired world are not likely to be evenly distributed.

● How internet can excite young to learn?

Internet can learn younger to learn much different new knowledge when they research any questions and find answers from internet channel.
As one major aspect of teen life is social environment, changes in how teens connect impact the ways in which teens develop social skills. ** Luckily, the Internet offers many social-skill enhancement opportunities for teens of all different personalities . One advantage the Internet brings that the standard school environment cannot is the ability for teens to adjust their amount of social interaction. Teens who are extremely outgoing can spend their free time in social environments both offline and online, making new connections and catching up with friends.For example, a teen who finds large amounts of face-to-face interaction to be intimidating can use the Internet to engage in conversations while reducing the potential for social anxiety. In a way, this trains less social teens to be more social . In the past, these types of teens did not have the advantage of this social training provided by the Internet.

● Internet can encourage Social Network Growth

The Internet offers teens the ability to make friends with peers with whom they would not otherwise connect. With pop culture deteriorating into many distinct subcultures, teens' interests are more variable than they have ever been. Whereas in the past, children at school might have discussed the current top 40 when discussing music, today's kids define their musical tastes as specific genres, such as post-industrial, dubstep or jpop. Today, it's harder for teens to find peers who share the same interests in their schools. But online, not so. The Internet's social networks help teens find communities of peers who share similar interests, allowing a teen to grow his social network in a way that is specific to him 2. Today's teens are increasingly willing to make friends with different groups of people due to the ability to actually meet them, and this can be useful when they reach adulthood, a time in which accepting people of different backgrounds and demographics is crucial to career and academic growth. The Internet offers teens the ability to make friends with peers with whom they would not otherwise connect.

Today's teens are increasingly willing to make friends with different groups of people due to the ability to actually meet them, and this can be useful when they reach adulthood, a time in which accepting people of different backgrounds and demographics is crucial to career and academic growth.But the Internet can help teens foster self identity through exposure to new people, communities, hobbies and concepts. As teens go through more experiences, they learn more about themselves. And as the Internet can offer teens a wealth of experience, it can play the role of hastening the development of self identity.For many teens, the hardest part of life is figuring out identity.But the Internet can help teens foster self identity through exposure to new people, communities, hobbies and concepts.

● What Are Main Benefits of Internet Communication speed improvement ?

It may include as below:

1 Makes communication easier

Doing business through phone or mail doesn't work well ? Before the internet came into existence, the only way to communicate was through a phone. Or if you needed to send a note you had to send letters via mail. With the arrival of the Internet, staff and team managers can connect instantaneously without leaving their work place. ezTalks Meetings, a one-stop internet communication provider, is a perfect example. With this platform, participants can communicate as if they were right next to one another thanks to its quality video and audio. The tool comes with a rich set of features like screen sharing, cross platform chat, innovative whiteboard, and more.

2 Enhances collaboration

Internet communication brings teams together across the globe. Staff can collaborate easily without limitations and make more informed decisions instantaneously. This leads to reduced project timelines, cutting back on the time required to launch a new product/service. This piece of technology is also useful in education. Not only can students collaborate with foreign students, they can share ideas and learn about the diverse cultures out there. Parents can also become actively involved in their kids education by linking their children school with libraries, homes, and more. Millions of schools around the world are already using this technology to enhance learning.

3 It is cost effective

The cost of internet communication is significantly low when compared with other means of communication like face to face meetings and mail delivery. The technology connects you to your partners, colleagues, clients and suppliers from just about any location for a fraction of the cost required to host a one-on-one meeting. And as technology continues to become more efficient, the cost of online communication continues to drop significantly. With the traditional face to face meeting, you need to spare time, cash to travel and so on. Internet communication allows you and your team to connect without having to leave your offices.

4 Improves work relationships

Building a good relationship between workers spread around the globe is not easy. Business trips can negatively affect life– work balance. Team members can burn out fast if they have to make business travels that deny them the chance to participate in crucial events with friends and family. With internet communication, employees can effortlessly communicate with one another at anytime from anywhere in the world. This allows employees situated in different parts of the world to give their opinion and voice their concerns. Therefore, internet communication is an important business asset, particularly for companies that have tapped into global markets.

5 Increases productivity

While the companies of yesteryear might not have treasured effective communication, modern workplace requires both the management and the staff have the tools to effectively communicate internally and externally. This is because effective communication is important in increasing productivity as it directly impacts the behavior of the employees and how they perform. Internet communication plays an integral role in getting stuff done fast and efficiently which ultimately improves productivity. Poor communication can have a negative effect on productivity as the staff may not get the adequate info to accomplish a job they have been assigned.

6 Increases accountability

Errors slow down productivity and so it is tempting to punish or fire employees who repeatedly make errors. One major advantage of internet communication is that it helps to decrease these errors. This piece of technology pinpoints errors and how staff can avoid them. In workplaces that don't make use of various forms of internet communication, those mistakes go unnoticed. With internet communication, there is no room for mistakes as employees feel liable for their actions and safe to point out mistakes. They also feel secure expressing their ideas and suggestions in a group setting.

● Why does internet improvement can help any industries services or efficiencies improvment?

Internet improvement will revolutionize the world and lead to groundbreaking changes in transportation, industry, communication, education, energy, health care, communication, entertainment, government, warfare and even basic research. For example, self-driving cars, trains, semi-trucks, ships and airplanes will mean that goods and people can be transported farther, faster and with less energy and with massively fewer vehicles. Automated mining and manufacturing will further reduce the need for human workers to engage in rote work. Machine language translation will finally close the language barrier, while digital tutors, teachers and personal assistants with human qualities will make everything from learning new subjects to booking salon appointments faster and easier. For businesses, automated secretaries, salespeople, waiters, waitress, baristas and customer support personnel will lead to cost savings, efficiency gains and improved customer experiences. Socially, individuals will be able to find AI pets, friends and even therapists who can provide the love and emotional support that many people so desperately want. Entertainment will become far more interactive, as immersive AI experiences come to supplement traditional passive forms of media. Energy generation and health care will vastly improve with the addition of powerful AI tools that can take a systems-level view of operations and locate opportunities to gain efficiencies in design and operation. AI-driven robotics (e.g., drones) will revolutionize warfare. Finally, intelligent AI will contribute immensely to basic research and likely begin to create scientific discoveries of its own. So, it implies that internet improvement ought assist any kinds of industy service or efficiency improvement.

● Internet may become any organizational digital assets

On an individual basis, we will think about our digital assets as much as our physical ones. Ideally, we will have more transparent control over our data, and the ability to understand where it resides and exchange it for value – negotiating with the platform companies that are now in a winner-take-all position. Some children born today are named with search engine-optimization in mind; we'll be thinking more comprehensively about a set of rights and responsibilities of personal data that children are born with. Governments will have a higher level of regulation and protection of individual data. On an individual level, there will be greater integration of technology with our physical selves. For example, I can see devices that augment hearing and vision, and that enable greater access to data through our physical selves. Hard for me to picture what that looks like, but 50 years is a lot of time to figure it out. On a societal level, AI will have affected many jobs. Not only the truck drivers and the factory workers, but professions that have been largely unassailable – law, medicine – will have gone through a painful transformation. It seems entirely reasonable that a great deal of our digital lives will be focused on habitable environments: identifying them, improving them, expanding them.

Significant, often highly communication and computation technologically driven, advances in day-to-day areas like health care, safety and human services, will continue to have a significant measurable improvement in many lives, often 'invisible' as an unnoticed reduction in bad outcomes, will continue to reduce the incidence of human-scale disasters. Advances in opportunities for self-actualisation through education, community and creative work will continue. So, I believe that future many organizations may apply internet communication tool for their digital assets.

● Internet improvement may assist robotic development

Most of the focus on technology and particularly AI and machine learning developments these days is limited to virtual systems (e.g., apps for travel booking, social networks, search engines, games). I expect this to move, in the next 50 years, into networking people with machines, remotely operating in a myriad of environments, such as homes, hospitals, factories, sport arenas and so on. This will change work as we know it today, as it will change medicine (increasing remote surgery), travel (autonomous and remotely-guided cars, trains, planes), entertainment (games where real robots, instead of virtual agents, evolve in real scenarios). These are just a few ideas/scenarios. Many more, difficult to anticipate today, will appear. They will bring further challenges on privacy, security and safety, which everyone should be closely watching and monitoring. Beyond current discussions on privacy problems concerning 'virtual world' apps, we need to consider that 'real world' apps may enhance many of those problems, as they interact physically and/or in proximity with humans. So, future historians will observe that, in many ways, the rise of the internet over the next few decades will have improved the world, but it hasn't been without its costs that were sometimes severe and disruptive to entire industries and nations as well as improve robotic development.

This is similarly valid for AI.Living longer and better lives is the shining promise of the digital age. Many respondents to this canvassing agreed that internet advancement is likely to lead to better human-health outcomes, although perhaps not for everyone. As the following comments show, experts foresee new cures for chronic illnesses, rapid advancement in biotechnology and expanded access to care thanks to the development of better telehealth systems. Life will improve in multiple ways. One in particular I think worth mentioning will be improvements in health care in three distinct ways. One is significantly better medical technology related to cancer and other major diseases. The second is significantly reduced cost of health care. The third is much higher and broader availability of high-quality health care, thereby reducing the differences in outcomes between wealthy and poor citizens. So, when hospitals can improve internet communication , if the hospital can apply robots to assist doctors and nurses to serve patients. Then, internet communication can help them to cooperate more efficient.

● Internet improvement to assist 5G laptop development

Many of the technologies we see commercialized today began in government and university research labs. Fifty years ago, computers were the size of walk-in closets, and the notion of personal computers was laughable to most people. Today we're facing another shift, from personal and mobile to ambient computing. We're also seeing a huge amount of research in the areas of prosthetics, neuroscience and other technologies intended to translate brain activity into physical form. All discussion of transhumanism aside, there are very real current and future applications for technology 'implants' and prosthetics that will be able to aid mobility, memory, even intelligence, and other physical and neurological functions. And, as nearly always happens, the technology is far ahead of our understanding of the human implications. Will these technologies be available to all, or just to a privileged class? What happens to the data? Will it be 'willed' as a digital legacy to future generations? What are the ethical (and for some, religious and spiritual) implications of changing the human body with technology? In many ways, these are not new questions. We've used technology to augment the physical form since the first caveman picked up a walking stick. But the key here will be to focus as much (or more) on the way we use these technologies as we do on inventing them. All of above factors will be influenced to future 5G mobile phone by internet improvement?

Our homes, transportation, appliances, communication devices and even our clothes will be constantly communicating as part of a digital network. We have enough pieces of this today that we can somewhat imagine what it will be like. Through our clothes, doctors can monitor in real time our vital signs, metabolic condition and markers relevant to specific diseases. Parents will have real-time information about young children. The difference in the future will be the constant sharing of information, data updates and responses of all these interconnected devices. The things we create will interact with us to protect us. Our notions of privacy and even liability will be redefined. Lowering the cost and increasing the effectiveness of health care will require sharing information about how our bodies are functioning. Those who opt out may have to accept palliative hospice care over active treatment. Not keeping track of children real-time may be considered a form of child neglect. Digital will do more than connect our things to each other – it will invade our bodies. Advances in prosthetics, replacement organs and implants will turn our bodies into digital devices. This will create a host of new issues, including defining 'human' and where the line exists between that human and the digital universe – if people are always connected, always on are humans now part of the internet?

● How internet improvement influences AI provides medical service to hospitals?

Similarly, AI embedded in devices or wearables can be applied to predict and ameliorate many mental health illnesses. However, there is potential for there to be huge inequalities in our societies in the ability of individuals to access such technologies, causing both social disruption and new causes for mental health diseases, such as depression and anxiety. On balance, I am an optimist about the ability of human beings to adjust and develop new ethical norms for dealing with such issues.Surveillance technology, especially that powered by AI algorithms, is becoming more powerful and all-present than ever before. But to look at that and say that technology won't help people is absurd. Medical technology, technology to help people with disabilities, technology that will increase our comfort and abilities as humans will continue to appear and develop.The digital revolution will bring benefits in particular for health, providing personalized monitoring through Internet of Things and wearable devices. The AI will analyze those data in order to provide personalized medicine solutions.The most noticeable change for better in the next 50 years will be in health and average life expectancy. At this pace, and, taking into account the developments in digital technologies, I hope that several discoveries will reduce the risk of death, such as cancer or even death by road accident. New drugs could be developed, increasing the active work age and possibility maintaining the sustainability of countries' social health care and retirement funds. Another area AI can have impact is in creating the framework within genomics, epigenomics and metabolomics can be used to keep people healthy and to intervene when we start to deviate from health. Indeed, with AI we may be able to hack the brain and other secreting cells so that we can auto-generate lifesaving medicines, block unwanted biological processes (e.g., cancer), and coupled to understanding the brain, be able to hack at neurological disorders."

Thus, I believe that future hospitals were able to utilize internet technology to solve human health problems to make citizens' lives better and improve their access to care and services to improve their health outcomes. The benefits of the internet in the health care industry have continued to improve access to care and services, particularly for elderly, disabled or rural citizens. Digital tools will continue to be integrated into daily life to help the most vulnerable and isolated who need services, care and support. With laws supporting these groups, benefits in these areas will continue and expand to include behavioral health and resources for this group and for others. In the area of behavioral health in particular, digital tools will provide far-reaching benefits to citizens who need services but do not access them directly in person. Access to behavioral health will increase significantly in the next 50 years as a result of more enhanced and widely available digital tools made available to practitioners for delivering care to vulnerable populations, and by minimizing the stigma of accessing this type of care in person. It is a more affordable, personalized and continuous way of providing this type of care that is also more likely to attain adherence.

● The cyborg generation: Humans will partner more directly with technology when internet is popular to be used in any where

The inevitable 'Singularity' will result in changes to humans and will increase the rate of our evolution toward hybrid 'machines.' I also believe that new and modified materials will become 'smart.' For instance, new materials will be 'self-aware' and will be able to communicate problems in order to avoid failure. Ultimately, these materials will become 'self-healing' and will be able to harness raw materials to manufacture replacement parts in situ. All these materials, and the things built with them will participate in the connected world. We will see continued blurring of the line between 'real' and 'virtual' life." For exaple, artificial general intelligence and quantum computing available in a future version of the cloud connected to individual brain augmentation could make us augmented geniuses, inventing our daily lives in a self-actualization economy as the conscious-technology civilization evolves. Implants in humans that continuously connect them to the web will lead to a loss of privacy and the potential for thought control, decline in autonomy.

● Everyone agrees that the world will be putting AI to work, when internet is improvement to raise robotic efficiency and performance improvement

The technology visionaries surveyed described a much different work environment from the current one. They say remote work arrangements are likely to be the rule, rather than the exception, and virtual assistants will handle many of the mundane and unpleasant tasks currently performed by humans. The shooting is done by a drone guided by a smart guy/gal working a 9-to-5 job in an air-conditioned office in a nice town. Garbage could be picked up, sorted, recycled, all by robots with AI. Tedious surgery completed by robots and teaching via YouTube would leave the humans to the interesting and exciting cases, not the redoing of same lessons to yet more patients/students. Humans could live well on a 20-hour work week with many weeks of paid vacation. Having a job/career could become a positive, not just a necessity. With 24/7 learning and just-in-time capacity, people could change areas or careers many times with ease whenever they become bored. This positive outcome is possible if we collectively manage the creation and distribution of the tools and access to the use of new emerging tools. Thus, future everyone will have hundreds of digital workers working for them. Our cognitive mediators will know us in some ways better than we know ourselves. Better episodic memories and large numbers of digital workers will allow expanded entrepreneurship, lifelong learning and focus on transformation.

Thus, our future social development already small world will shrink further as remote collaboration becomes the norm, resulting in major social changes, among them allowing the recent concentration of expertise in major cities to relax and reducing the relevance of national borders. Furthermore, deep learning and AI-assisted technologies for software development and verification, combined with more abstract primitives for executing software in the cloud, will enable even those not trained as software engineers to precisely describe and solve complex problems. I believe the question we're facing is not 'When will machines surpass human intelligence?' but instead 'How can humans work together with machines in new ways?' Rather than worrying about an impending Singularity, I propose the concept of Multiplicity: where diverse combinations of people and machines work together to solve problems and innovate. In analogy with the 1910 High School Movement that was spurred by advances in farm automation, I propose a 'Multiplicity Movement' to evolve the way we learn to emphasize the uniquely human skills that AI and robots cannot replicate: creativity, curiosity, imagination, empathy, human communication, diversity and innovation. AI systems can provide universal access to sophisticated adaptive testing and exercises to discover the unique strengths of each student and to help each student amplify his or her strengths. AI systems could support continuous learning for students of all ages and abilities. Rather than discouraging the human workers of the world with threats of an impending Singularity, let's focus on Multiplicity where advances in AI and robots can inspire us to think deeply about the kind of work we really want to do, how we can change the way we learn and how we might embrace diversity to create myriad new partnerships. So, future

AI and internet technoloy will become new partners to assist any business development, even any organizations and social development. Hence, internet improvement must be needed in order to let any businesses can apply robots to raise efficiencies and improve performance more effectively. For example, free internet-connected devices will be available to the poor in exchange for carrying around a sensor that records traffic speed, environmental quality, detailed usage logs, and video and audio recordings (depending on state law). There will be secure vote-by-internet capabilities, through credit card or passport verification, with other secure kiosks available at public facilities (police stations, libraries, fire stations and post offices, should those continue to exist in their current form). Internet and 24/7 real-time connectivity will no longer be viewed as a 'thing' independent from daily life, but integral, like electricity. This has profound psychological implications about what people assume as normal and establishes baseline expectations for access, response times and personalization of functions and information. Contrary to many concerns, as technology becomes more sophisticated, it will ultimately support the primary human drives of social connectedness and agency. As we have seen with social media, first adoption is noncritical – it is a shiny penny for exploration. Then people start making judgments about the value-add based on their own goals and technology companies adapt by designing for more value to the user . Technology is going to change whether we like it or not – expecting it to be worse for individuals means that we look for what's wrong. Expecting it to be better means we look for the strengths and what works and work toward that goal. Technology gives individuals more control – a fundamental human need and a prerequisite to participatory citizenship and collective agency. The danger is that we are so distracted by technology that we forget that digital life is an extension of the offline world and demands the same critical, moral and ethical thinking.

In future 50 years every aspect of our life will be connected, organized and hence, partly controlled, as technology platform and applications businesses will take this opportunity. A few global players will dominate the business; smaller companies (startups) will mostly have a chance in the development sector. Many institutions, such as libraries, will disappear – there might be one or two libraries that function as museums to show how it used to be. People who experienced today's world will definitely value the benefits and amenities they have through technology (human-machine/AI collaboration). If technology becomes part of every aspect of our lives we will have to give up some power and control. People thinking in today's terms will lose a certain amount of freedom, independency and control over their lives. People born after 2030 will probably just think these technologies produced changes that are mostly for the better. It has always been like this – people have always thought/said 'in the old days everything was better. The free, open internet that represented a set of decentralized connections between idiosyncratic actors will be recognized as an aberration in the history of the internet. Today's internet giants will probably be the internet giants of 50 years from now. In recent years, they've made substantial progress in curtailing innovation through acquisitions and copying. As the industry matures, they will add regulatory capture to their skill sets. For many people around the world, the internet will be a set of narrow portals where they exchange their data for a curtailed set of communication, information and consumer services. Thus, digital tools will be part of our body inside and remotely, and will assist us in decision- making constantly, so it will become second nature. Nonetheless, physical feelings will still be exclusively 'physical,' i.e., there will be a significant difference between the 'sensor-based feelings' and real body feelings, so human beings will still have some advantages over technology. This, I believe, will last forever.

CHAPTER VI

The relationship between robotic invention and economic growth

Can robotic invention bring social economic growth? If it is possible that why and how robotic invention can assist any countries social economic growth? First, we need to know whether what economic growth means in order to answer this question. In macro economic view, economic growth may mean that GDP growth, employment ratio growth, job creating growth, unemployment ratio reduces, consumption growth, productive industries growth, service provision and service needs increases. So, it seems that any countries' social development , when it can have positive impact growth for any one of these issue. It implies that the country's economic is growing.

To discuss AI and economic growth issue, it may being these two main questions. Whether robotic invention may have direct or indirect relationship to influence any countries' economic growth? What are the main factors to cause robotics invention to bring the country's economic growth in its society?

On the first hand, some researchers believe that robotic invention may influence any countries' economic growth. However, otherwise, other many researchers find large and robust negative effects robots on employment and wages. They estimate what are more robot per thousand workers reduces the employment -to-population rate by between 0.18 and 0.34 percentage points, and is associated with a wage decline of between 0.25 and 0.5 percentage. Because many jobs can be replaced by robotics. So, it seems that robotics invention can reduce workers number and increasing wage level, even robotic increasing number it can influence unemployment ratio increases to the country. But, some lecturers estimate that it can impact economic growth. They estimate that AI may deliver an additional economic output of around US$13 trillion by 2030 year, increasing global GDP by about 1.2% annually. This will mainly come from substitution of labour by automation and increased innovation in products and services.

What is the impact of robots on society? They may include this spillover, one robot per thousand workers has slightly less of an impact on the population as a whole, leading to an overall 0.2% point reduction in the employment-to-population ratio, and reducing wages by 0.42%. Thus, adding one robot reduces employment nationwide by 3.3 workers. So, it seems that robotic number increases may influence global workers number increases many influence global workers number reduces as the same time. It can cause unemployment workers number increases in societies.

On the other side, robotic invention can increase productivity, because robots increase productivity, which means that fewer human hours are needed to produce a given output. But, higher productivity also reduces production costs and output prices. Consequently, robotic production anticipation , it can increase the quality demanded by consumers, and firms hire workers in this increased demand.

But, when robotic production participation to any industrial manufacture, it can also bring negating effects, when they are entering the workforce, e.g. higher maintenance and installation costs to factories, enhanced risk of data breach and other cybersecurity issues, reduced flexibility, anxiety and insecurity regarding the future social development. So, it seems that robotic invention can bring the future of workplace automation may reduce workers number of factories, loss of jobs and reduced employment opportunities to global future workers, potential job lose, initial investment costs to employers. However, AI may also bring harmful to our future societies because if AI surpasses humanity in general intelligence and becomes " superintelligent" , then it could become difficult or impossible for humans to control. Moreover, a second source concern is that a sudden and unexpected " intelligence explosion" might take an unprepared human race by surprise. Hence, robotic invention may become human enemy or soldier, if human applies it to social damage aspect.

However, robotic invention can also bring advantages to our future societies in possible, robotic automation may bring advantages to employers : cost effectiveness, improved quality assurance, increased productivity, avoiding workers need to work in hazardous environments. But,. AI can also bring positive impact our daily lives, such as artificial intelligence can dramatically improve the efficiencies of our workplace and can argument the work humans can do. When AI takes over repetitive or dangerous tasks, it frees up the human workforce to do work, they are better equipped for, tasks that involve creativity and empathy among others.

However, robotic invention may bring organizational benefits in our societies. Robots will have a profound effect on the workplace of the future. They will become capable of taking on multiple roles in organizations, e.g bookkeeping or writing law draft simple clerical tasks. So, it's time for us to start thinking about the way we shall interact with our new coworkers. To be more precise, robots are expected to take over half of all low-skilled jobs in our societies, e.g. cleaning , warehouse picking up delivering tasks, restaurant cooking, hotel front line customer service, hotel room food delivery tasks etc.

So, when robots would be used in many fields all over the world. However, robots can not totally rule over the workplace by replacing all humans at jobs to keep economy afloat. Hence, robots in the workplace may bring advantages, they will not have bad emotion problems, such as safety of utilizing robotics to work in dangerous workplace. Robots do not get distracted or need to take breaks robots never need to sleep or they need to divide their attention between a multitude of things, perfection, let employees to feel happier and safe to work when robotics can help workers to work in dangerous warehouses or factories, workers can be replaced from robotics, increases productivities and job creation , even raises efficiencies in any workplaces.

In our future, whether robots won't destroy humans. It depends on how humans choose to apply this technological worker tool. A robot may not injure a human being or, though inaction, allow a human bring to come to harm. A robot must obey the orders given it by human beings, expect where such orders would conflict with the first law. A robot must protect its existence as long as such protection does not conflict with the first or second laws. How AI technology affects us in the future. They are concerned that we will see increases in stress, anxiety, and depression as digital lives expand. Meanwhile , we shall need to adapt our future digital living, there will be less face-to-face interaction , increased inactivity, poor in-person communication skills and an overall distrust among people. For robotic invention case example, future robotics can may focus on developing these five major field: Human -robotic interface, mobility, manipulation, programming, sensors and their importance to robotics development.

Robots can be applied to educational aspect. Robots can be used to bring students into the classroom that otherwise might not be able to attend. Robots such as the one mentioned are able to bring school to student who can not present physically. Simulators-high school sees the strongest example of stimulators within drivers' education courses, e.g. Google's worker robots. Google is planning to produce worker robots with personalities. So, robots can help teachers automate key classroom processes, integrate advanced technologies, acclimate students to technological change and helping identify personalized learning potential and discovering key learning trends.

● How manufacturers to raise efficiency in order to improve economic growth in possible to apply manufacturing robots to improve help economic growth

(AI) can be defined as the capability of a machine to imitate intelligent human behavior. If AI can imitate any talent humans to learn how to improve their behaviors, e.g. manufacturing industry worker behavior improves to raise GDP growth or productive number grows rapidly . Can artificial intelligence being labour to become automated? Allows an ever-in-to increasing number of tasks previously performed by human labour to become automated in the ordinary production of goods and services process.

Can (AI) create new ideas and technologies to help businessmen to solve complex problems and improve automation in the production of goods and services. The question concerns how (AI) manufacturing technology can impact economic growth?

(AI)'s new form of automation live, self-driving cars, or they may bring high levels of skill,such as legal services, radiology, and some forms of scientific lab-based research. Can it allow our societies be impacted on economic growth, due to automation to disicipline our modeling of AI.

In fact, AI automation to production of new ideas to future any industrial manufacturing process. It can influence to market structure, organization restructure, reallocation and wage inequality. So, discovering to AI automation can help future any organizations need to change existing task or discovering new tasks that can be used in production a reflects the fraction of tasks that have been automated.

It brings automating old tasks when automated could be constant, leading a stable , capital share and a stable growth rate. Hence, in long term, AI automation manufacturing technology may help businessmen to reduce large manufacturing cost in order to achieve stable micro economic benefit to them when they can apply AI automation manufacturing technologies to improve their productivities in efficient manufacturing method. For Coca Cora soft drink example, if Coca Cora applies AI automation to raise its productive soft drinks number. Then , it can manufacture many soft drinks number in short time per day. Then, its sale number can be increased, when it has enough number to supply to global to sell its soft drink. Consequently, soft drink sale number must grow rapidly significantly.

The fact that automated goods are produced with cheap capital , but it can also help business to raise production number significantly . How this superintelligence affect the economy? It seems physical tasks are essential to producing output, but when the manufacturer applies robotics to help production . Then, employees number may reduce, due to robotic participation, wages expenditure may reduce , but production number may increase .

So, AI increases the motivation at physical tasks. Hence, AI must may bring production growth innovation incentives to any manufacturers. Finally, with imitation and learning being performed mainly by super machine in developed economies. Then , research labor would become devoted to product innovations increasing product variety or inventing new products (new product lines), to replace existing products.

It is one good example to explain that how AI can bring long term social economic benefit to any manufacturers, when any old (existing) product lines are improved to innovative new product line by robotic manufacturing participation. Moreover, AI can change market structure to be reduced competition. When be escape competition effect tends to dominate at low discouragement effect may dominate for higher levels of competition or in less advanced economies. Hence , AI can also affect innovation and growth through potential effects, it might have on product market competition. So, it seems that (AI) can respond on helping social economic growth principle. On conclusion, although robotic invention may bring disadvantages to raise unemployment ratio in possible when there are many low-skilled jobs are replaced by robotics, but at the same time, robotic may be applied to education aspect, when we are experiencing digital knowledge social development stage. Hence, smart robots ought may help humans to raise economic long term growth in possible when robotics can help the developing countries to develop more rapid to be developed countries as well as they can help the developed countries to develop more advanced both in global whole one economic developed societies. So, when robots can assist any countries to cooperate together, any countries are not independent, we need robots to assist develop between countries. Then, robots can assist any countries to develop economic growth in possible in future this day comes.

Chapter 8

How robotic helps to solve recession

Is the use of robots to job market increasing during the Great Recession? Some researchers constructed a measure of the use of robots—commonly referred to as "robot intensity"—to estimate trends in robot exposure across more than 250 metropolitan areas and over time, finding that: During the Great Recession, robot intensity plummeted. But since 2009, robot intensity has sharply increased nationwide. They felt that robots may influence recession in possible. How are robots going to affect our jobs? Most analysis tends to be prospective in nature, and estimates of future impacts on employment vary widely, with some studies predicting that as many as 50 percent of all workers are at risk of losing their jobs to automation. Even less is understood about the actual impacts of robots on jobs, wages, and workers today. If there are many low skillful jobs e.g. cleaner, warehouse deliver, cooker, restaurant waitor etc., even high skillful jobs, e.g. accountant, lawyer, doctors etc. occupations are replaced by robotics. Then, our societies will increase unemployed people number. Consequently, great recession will be caused by robotic workers because when our societies have many people lose jobs, then many people loss income, then our consumption desires may be influenced to reduce. Consequently, many businesses may lose many customers. Low consumption desires may bring serious recession to any countries, due to robotic workers number increases to replace human workers in global societies.

The reason is that new technologies of the period have enabled people to be very productive while working part-time. Businesses do not need large numbers of employees, so individuals can devote most of their waking hours to hobbies, volunteering, and community service. In conjunction with periodic work stints, they have time to pursue new skills and personal identities that are independent of their jobs. Developed countries may be on the verge of a similar transition. Robotics and machine learning have improved productivity and enhanced the economies of many nations. Artificial intelligence (AI) has advanced into finance, transportation, defense, and energy management. The internet of things (IoT) is facilitated by high-speed networks and remote sensors to connect people and businesses. In all of this, there is a possibility of a new robotic society that could improve the lives of many people, but it also encourage future businesses apply robotics to replace human workers to do many jobs in finance, transportation, defense and energy management, medical health, hotel, tourism, cinema, theatre etc. entertainment service fields. So, robotics may bring reducing cost benefits to businessmen, but they can also increase workers losing jobs number in our societies if future many businesses make decision to apply robotics to replace human workers to do any simple ot complex jobs in global job market.

A McKinsey Global Institute analysis of 750 jobs concluded that "45% of paid activities could be automated using 'currently demonstrated technologies' and . . . 60% of occupations could have 30% or more of their processes automated."[6] A more recent McKinsey report, "Jobs Lost, Jobs Gained," found that 30 percent of "work activities" could be automated by 2030 and up to 375 million workers worldwide could be affected by emerging technologies.

Researchers at the Organization for Economic Cooperation and Development (OECD) focused on "tasks" as opposed to "jobs" and found fewer job losses. Using task-related data from 32 OECD countries, they estimated that 14 percent of jobs are highly automatable and another 32 have a significant risk of automation. Although their job loss estimates are below those of other experts, they concluded that "low qualified workers are likely to bear the brunt of the adjustment costs as the automatibility of their jobs is higher compared to highly qualified workers."

reference

James Manyika, Susan Lund, Michael Chui, Macques Bughin, Jonathan Woetzel, Parul Batra, Ryan Ko, and Saurabh Sanghui, "Jobs Lost, Jobs Gained: Workforce Transitions in a Time of Automation," McKinsey Global Institute, December, 2017.

Melanie Arntz, Terry Gregory, and Ulrich Zierahn, "The Risk of Automation for Jobs in OECD Countries," Organization for Economic Cooperation and Development, Working Paper 189, 2016.

However, some economists felt opposite opinions, they believes that future many businesses won't choose whole applying robotics to replace human workers in global job market. So, they are only human workers assistant role. Economists have, on the whole, been fairly discuss about the impact of robots and AI on workers. History is strewn with incorrect predictions of the looming irrelevance of human labour. The economic statistics have yet to signal the arrival of a robot-powered job apocalypse. Outside of slumps, firms remain keen to hire humans, for example. Growth in productivity—which ought to be surging if machines are helping fewer workers produce more output—has been unimpressive. A look beneath the aggregate numbers, though, reveals that change is indeed afoot. They believes that an AI-induced change in the mix of jobs need not translate into less hiring overall. If new technologies largely assist current workers or boost productivity by enough to spark expansion, then more AI might well go hand-in-hand with more employment. This does not appear to be happening. Instead the authors find that firms with more AI-vulnerable jobs have done much less hiring on net; that was especially the case in 2014-18, when AI-related vacancies in the database surged. But the relationship between greater use of AI and reduced hiring that is present at the firm level does not show up in aggregate data, the authors note. Machines are not yet depressing labour demand across the economy as a whole. As machines become cleverer, however, that could change.

Take work by Daron Acemoglu and David Autor of the Massachusetts Institute of Technology, Jonathon Hazell of Princeton University and Pascual Restrepo of Boston University, which was presented at the recent meeting of the American Economic Association (AEA). The authors use rich data provided by Burning Glass Technologies, a software company that maintains and analyses fine-grained job information gleaned from 40,000 firms. They identify tasks and jobs in the dataset that could be done by AI today (and are therefore vulnerable to displacement). Unsurprisingly, the researchers find that businesses that are well-suited to the adoption of AI are indeed hiring people with AI expertise. Since 2010 there has been substantial growth in the number of AI-related job vacancies advertised by firms with lots of AI-vulnerable jobs. At the same time, there has been a sharp decline in these firms' demand for capabilities that compete with those of existing AI.

An AI-induced change in the mix of jobs need not translate into less hiring overall. If new technologies largely assist current workers or boost productivity by enough to spark expansion, then more AI might well go hand-in-hand with more employment. This does not appear to be happening. Instead the authors find that firms with more AI-vulnerable jobs have done much less hiring on net; that was especially the case in 2014-18, when AI-related vacancies in the database surged. But the relationship between greater use of AI and reduced hiring that is present at the firm level does not show up in aggregate data, the authors note. Machines are not yet depressing labour demand across the economy as a whole. As machines become cleverer, however, that could change.

Evidence that AI affects labour markets primarily by taking over human tasks is at odds with some earlier studies of how firms use the technology. A paper from 2019 by Timothy Bresnahan of Stanford University argues that the most valuable applications of AI have nothing to do with displacing humans. Rather, they are examples of "capital deepening", or the accumulation of more and better capital per worker, in very specific contexts, such as the matching algorithms used by Amazon and Google to offer better product recommendations and ads to users. To the extent that AI leads to disruption, it is at a "system level", says Mr Bresnahan—as Amazon's sales displace those of other firms, say.

New work by Ajay Agrawal, Joshua Gans and Avi Goldfarb of the University of Toronto suggests that this state of affairs may not persist for long, though. As the quality of AI predictions improves, they write, it becomes increasingly attractive for AI-using firms to restructure in more radical ways. At some level of accuracy, for example, Amazon's ability to predict consumers' desires could encourage the firm to adjust its business model—by pre-emptively shipping goods to consumers before they ever go searching at Amazon in the first place—in ways that are likely to change how many workers and of what sort the firm requires. In that event, the influence of AI on the economy could change dramatically. So, such as Amazon case, it applies robotics are only concentrated on predicting consumer prediction aspect, AI is only assistant role to Amazon human market researchers. They help Amazon market researchers to gather consumer behavior data , but Amazon human market researchers need to do marketing analysis tasks by themselves. So, Amazon can not employ human market reseachers jobs position in itself company. Amazon needs robotic and human market researchers to do market research tasks in order to achieve how to predict consumer behavior more accurately. Thus, it seems that future large enterprises won't fire any professional staffs more easily because some complex tasks, e.g. analysis tasks, they believe that human's analysis can make more accurate judgement to compare AI's analysis.

However, some scientists believe that some skilful professional occupations , they have possible be replaced by robotic. Will Architects and Engineers be Replaced by Robots? It's not uncommon for people to think they may be replaced by a robot in the workplace. After all, it's happened plenty of times before. For example, the rise of the mechanical assembly line saw machines replace people in the early 20th century. With recent advances in artificial intelligence (A.I.), it's entirely possible that more jobs are at risk.

Even skilled workers, such as architects, programmers and engineers may be at risk. One day, an A.I. software developer may be able to do everything that a human programmer can do.

Recent reports have not abated this thought process. In fact, the 2016 Economic Report of the President seemed to suggest that artificial intelligence is playing an increasingly important role in the engineering industry. Just think about the software you use in your work. Many software packages can handle a lot of the complex calculations for you. Yes, this cuts down on the amount of work you do. However, this automation may also present a threat to your job. What if the future sees these same software packages handling data input, as well as processing.

Automation is important. The use of artificial intelligence, alongside various other technologies, has always improved production. More work gets done, which means that businesses make more money. Architects and engineers constantly look for ways to speed up their work. The desire for automation has informed many recent software innovations. Furthermore, project methodologies, like Building Information Modelling, place automation at the fore. That's great for speed and efficiency, but what does it mean for architects and engineers? History has shown that automation has a very human effect. People lose their jobs because machines can do them faster. Just think of it from a business viewpoint. Do you want to pay 10 or more employees, or invest in one machine? More often than not, the machine will cost less than the employees, even if you factor maintenance into the equation. It's a simplification, but not an invalid one. Businesses make these sorts of decisions all the time. By pushing for automation, architects and engineers may be slowly working themselves out of their own jobs.

Several studies have also suggested that artificial intelligence may cause job losses. One recent example comes from the University of Oxford. The study found that over 700 types of jobs are at risk of technological disruption. All told, this means that about 47% percent of jobs are at risk because of artificial intelligence. That is a huge amount of people who may find themselves obsolete due to advancing technology. The same study also mentioned a concept called the "technological bottleneck". The researchers used this to determine how "at risk" a job was of displacement. The bottleneck takes three factors into account:

•How much creative intelligence the role needs

•If manual manipulation and perception is required

•The role of social intelligence in the role

If a role requires a high degree of any of those three things, it's less likely that it's at risk from artificial intelligence. Architects and engineers are a good example. These professionals require a great deal of creative intelligence. Artificial intelligence and robots may not be able to emulate that creative intelligence. As a result, it's unlikely that architects and engineers need to worry about losing their jobs. Right now, at least. The study concluded with a cautionary note. It said that just because automation enhances an architect and engineer's work right now, it doesn't mean that automation won't replace that role in the future. So, if future human architects or engineers jobs can be replaced to do by robotics. Then, robotic architects can help the architectural firms to design more attractive architectural plans to satisfy construction firms clients needs in short time or robotic engineers can help the engineering design firms to design more attractive machines to satisfy any engineeing customers needs in short time , when robotics can be invented to own excellent creative ability to compare human architects or engineers. So, these two professional occupations will be lost when robotic architects and robotic engineers can be invented to own excellent creative ability to compare human architects or engineer in future one day in possible.

However, robotic architects or engineers may bring advantages and disadvantages both aspects:

On advantages aspect:

Artificial intelligence allows us to do all of the following:

•The completion of mundane tasks that would otherwise take a lot of labour hours. Automating such tasks frees up skilled workers to work on more important tasks.

•A.I. is not as prone to making errors as a person. As long as the A.I.'s programming is good enough, you should find that calculating errors and similar issues become problems of the past.

•Speed is a key feature of artificial intelligence. Huge datasets no longer provide any problems to businesses, as automation allows for much faster processing. This means that a business can spend money elsewhere.

•The most complex A.I.s reduce the amount of risk attached to the decision-making process. The "Curiosity" Mars rover is a good example. It's programmed to choose the best course of action depending on its position.

On disadvantages aspect:

It's not all good, unfortunately. The following are some of the bad points of artificial intelligence:

•The previously mentioned job losses can cause all sorts of problems for staff morale.

•Some believe that artificial intelligence gets rid of the human element. The nightmare scenarios in films like "The Terminator" may seem far-flung, but that doesn't mean there isn't a risk in letting machines make all the decisions.

•A.I. relies on pre-existing knowledge, which means it lacks creativity. Attempting to use it for creative endeavours may result in failure.

•Algorithms may not be able to make judgement calls in disaster situations. Again, the A.I. may not take the human element into account, no matter what's actually happening on the ground.

Oe people management aspect, what do you think would be the reaction to a robot attempting to manage people? It's likely that a lot of people won't take to kindly to artificial intelligence telling them what to do. Many underestimate the importance of people skills in the architecture and engineering profession. Architects and engineers must be able to organise workloads and manage individuals. It is sure, A.I. device could handle the former. Scheduling is a task that many already automate. However, A.I. will fall down when it comes to the human relationships that are so vital in a team environment. An A.I. won't understand when somebody is demotivated, or why. It won't make allowances for the human issues that affect every problem. This makes skilled team members even more valuable. As A.I. takes an increasing role in the workplace, the need for people management will become more important. Architects and engineers with those skills may even find they make more money to employ them. Although, it is possible that AI can replace human architects or engineers to do their tasks to be better , it can help any one architectural or enginering firms to improve design performace in order to satisfy customers design demand, but our societies will increase unemployment ratio to engineers and architects number, even universities will reduce architect and engineering students number. Our traditional professional knowledge will be felt to be rubblish when these professional subjects won't be useful to help us to find jobs easily. So, AI invention will influence many students won't choose to study these two subjects. Our societies will be influence to experience knowledge recession when knowledge will become rubblish because robotics invention , it can do many human professional jobs to do. " knowledge recession" will be important factor to bring economic recession, because human can not be encouraged to learn any new knowledge to prepare to enter job market due to robotic invention can replace us to do more complex tasks in our future societies.

Can robotic leadership be good solution method when recession had come to the country?

On leadership management aspect, can robots become clever leadership to any organizations? As artificial intelligence becomes further embedded into our everyday working lives, we are already seeing the footprint of machine learning, automation, algorithms and robots in many of our professions and sectors. However, when we look at the upper levels of business management and leadership, these technological shifts are less evident, with C level Executives continuing to lead and strategise as they have done before. In Ireland, there are more than 500 CEOs. The question is, when will we start to see machines and robots play a more central role in the CEO sphere, and is a 'Robot CEO' realistic in the short to medium term?

New research shows that 24% of people aged 25-29 would replace their boss with a robot, demonstrating an interesting trend among Generation Z. However, these data sets are perhaps less founded in AI and robotics and more in current employee engagement. It's telling that the 20-30% of people who would willingly replace their human boss with a robot is about the same percentage of people who are consistently classified as "actively disengaged" at work. In addition, research from analytics giant Gallup demonstrates that 70% of how we feel about work, namely our emotional commitment, is driven by who our manager is, again underlining the centrality of human behavioural traits when making decisions on leadership.

As the Irish economy moves forward, values will define how we use and leverage the potential of AI. Tomo Noda of the Harvard Business Review believes that we will need more focus on leadership with humanity, ethics and integrity, stating "only good people can create good AI."With many roadblocks and challenges for the economy looming, primarily in the shape of Brexit and trade tariffs, it is a sound integration of both human and tech which will provide the leadership required to ensure our economy remains robust. Human leaders have played a central role in helping to steer us out of the 2008 recession, and with diplomacy and relationship building key to our post-Brexit future, humans will undoubtedly be the key influencers within the C level for decades to come. Hence, it seems that future organizations ought choose to apply robots to assist leaders to do make important decision, if robotics can assist leaders to make any important decision to conclude the best results to improve any companies performance. Then, GDP may be influenced to increase or grow rapidly, when recession had come to the country. So, robotic leadership may be one solution to solve recession method in possible.

The Recession Cometh and Robots are Ready

In economic demand vs. supply theory indicates that consumer appetites for customized product and their expectations for ever-lowering costs. So the current tug-of-war over if, when, and where a recession will hit is not unchartered territory. For manufacturers, though, the uncertainty is particularly challenging, as the flexibility that allows operations to reflect the pace of the economy simply isn't there. The economy has been growing. Unemployment is down. Last year's Christmas sales were better than they've been in a long time. All good and logical reasons for manufacturers to hire.

Recently though, there have been signs that instability is coming. The US stock market experienced extreme volatility as 2018 came to a close. The Federal Reserve raised interest rates and laid down some pretty clear language that more was to come. Consumer confidence fell. In the UK, a deal on Brexit that would allow British manufacturers to continue to do business with the EU seemed elusive at best. The Chinese government announced that growth in its economy has slowed. And the "R" word started to appear with more frequency. These are not signs that inspire confidence. So, manufacturers once again find themselves in a place they know so well. The rock: the need to hire workers to keep ahead of demand. Compounding this challenge is that unemployment is low and it is very hard to find people with the skills needed to take a job in manufacturing and be ready to work on day one. The hard place: overwhelmingly, today's automation is fixed, expensive, and able to perform only a single task.

As one supply chain executive of a global automotive firm shared recently, "In a downturn...it is about flexibility. All of the automation we have cost too much and it is too complicated to change what it does. What we need is flexible automation that can respond when and how we need it to."Can robotics be applied to manufacturing industry to avoid cost reduces to manufacturers when consumption number reduces or recession is coming? So what makes the most sense? Hire, hoping that if and when recession comes, it will be short-lived and you won't have to lay folks off? Or try to invest in reconfiguring existing automation?

Hence, cobots give manufacturers the flexibility they need to thrive in good times and not-so-good times. Advances in robotic technology make it possible to put cobots to work

•at lower costs

•on more than a single task

•in the same amount of time, it takes to train a person – or even less

With collaborative robots, manufacturers can build the operations they need to compete and thrive regardless of the economic climate, where manufacturing robotic participation can help organizations to reduce labours number on strategic tasks and flexibility is part of the organizational human resource cost reducing strategy. It seems that robotic manufacturing workers can help organizations to reduce manufacturing cost when recession is coming. Consequently, these applying manufacturing robotic businesses may prolong business life time in possible. So, it seems that manufacturing robots may help organizations to reduce manufacturing cost to keep life when recession is coming.

CHAPTER VII

The relationship between social change and human behavior

Human Behavioral network job brings social economic benefits

What does human network job mean ? Why may human network job be popular? Why human network job behavior may influence economy ?

Nowadays internet is popular to use. We can apply internet to find data , search any new things, even earn money. Why does internet

may become huma network job source. For example, e-publish may be one kind of new human network job. Any authors may apply internet

channel to help them to sell electronic or paper books from e-publisher web store. They may apply facebook, you tub etc. any online

channel to promote themselves new books to let new readers to know whether when they may buy themselves favourable new topic books to read from electronic publisher web store.

Thus, future electronic publisher industry may help any authors to build internet network platform to help them to sell and promote

ot advertise their any one new electronic or paper book topic to let global any one reader to choose to buy their any new topic books from electronic publisher web store easily and conveniently. However, it implies that electronic network platform author may be one kind of future new human network job in our societies.

How electronic network platform author job may bring economy benefit in macro economy view? A person can have few friends, contacts and still be very influential if these few

friends and contacts are themselves highly influential, e.g. one author must not need to know any one reader in global society. When they like to choose any electronic books from electronic internet network platform. They may become the author's any one topic book buyer, when they feel the author's any one topic book is fun and attract they make decision to buth the strange author whose the topic book from electronic book publisher's platform web store conveniently in short time. Although, they are strangers, they do not know themselves , but the reader can understand what it way that made Google from writing platofrm to create new creative mind and typing network job method to replace traditional hand writing book method for global authors. It will be one kind of new human network writing job.

Hence, global any one reader can apply an innovative search engine , such as google.com to find whether whom author personal new topic books are value to read from internet.

Then, the electroniuc publisher's web store may be new book store platform sale network to help the author to sell many electronic or paper books from electronic network platform

in short time. So, internet may be future new network plaform to help global any one author to create network writing job absolutely. Furthermore, internet may be popular social media

to help any one author to build goold relationship between his/her readers. It is one kind of new network, human network job. New authors do not need to buy many paper books to prepare to put in any one book shop warehouse. Their every book can print on demand to reduce out of book stock in any one book shop. They may choose to sell either electronic books or paper books both from any one book publisher web store. So, electronic network platform may be one kind of good writing channel to help human authors to create income and it can also help authors to bring new creative mind and new topic fun content books to let readers to know and buy to read from electronic publisher network platform.

Why does human behavior may be one kind of new human network job to bring global economic advantages. ALthough, it may be free income or without inocme, but the person does the network behavior, his/her behavior may be bring advantages to influence many other people's health. For this case, when a worker in a coffee shop in an airport gets a vaccination againnst the flu, it does not only helps him or her stay healthy, but also helps the many travellers who might otherwise have been inflected if that workers caught the flu. So, the externality , the result implies the vaccination of even a part of a community conveys benefits to the whole community. For example, governments pay special attention to the vaccinations of school children, teachers, health mothers, and the elderly, categories of people particularly susceptible not only to catching, but also to transmitting a disease.

It is not accidential that governments are heavily involved with vaccination . When there are externalities, free market, fail to persuade individual incentives with society's their the worker's decision of whether to get a vaccine ends up attracting whether other people get sick. The workers might not fully take all these other people's potential suffering into account when making her or his vaccination decision.

As Stanford University does many suggestions, understand this and tries to help them make the right decisions and so providers free flu vaccines for its staff and students.

Small pockets of unvaccinated individuals can allow a disease to gain a spread more widely well-being. For example, parent weighing the costs and benefits of a vaccine for their child is not always thinking of the consequences of that vaccination to other people. THese are markets in which subsidizing or regulating behavior can make everyone better off. Because the reason for requiring that a child be vaccinated before enrolling in school is not just to protect that child, because each child's vaccination affects others via potential contagions.

Robots take our jobs behavioral and economy influences

Robot job behavior brings economy influences

If one day robots can replace human to do simple, even complex jobs. They will bring what influences to our global societial economy.The popular economic refrain declares that the

global middle class is dying and robots will soon take our jobs, e.g. shopping center customer service jobs, library service jobs, cinema ticket sale jobs, restaurant kitchen cooker jobs,

even, bus drivers, taxi drivers etc. public transport driving jobs, accountant, doctors etc. professional jobs. Whether it is beautiful or petty matter if our future societies have many human jobs can be replaced to do from robots. Businessman must may reduce to employ employees and reduce to pay salary or wage, when robots can be replaced to do their employees tasks. But, societies must bring unemployement rate rises , due to societies will have many people loss jobs when their employers choose to buy robots to serve their clients or do any office tasks or customer service or cleaning etc. tasks.

In micro economy view, employers may save money in long term, but in macro economy view, it will cause unemployment ratio rises , even crime rate rises when there are many people lose

jobs in societies. These models of doom, though, fail to account for the hundreds of businesses riding the waves of change in their industries when robots may be invented to replace human to do many simple , even complex tasks in our future societies.

WE may image that one small factory needs to manufacture fishes canes to sell to supermarket, the small , cheaper stuff and higher margin parts of the fishes manufacture industry. Before, this factory needs to employe many human factory workers need to help every fresh customer makeing the perfect fishing gear, designed for performance, durability, and cost in order to achieve to manufacture every fish cane in whole fished processing manufacturing stages. Every worker needs to spend about 15 to twenty minutes to finish every fish cane , till to delivery to any supermaket to sell. If this fish canes manufacturing factory can apply manufacturing robots to help them to finish any one working tasks , every robot can only spend five minutes to finish whole fresh fish cane manufacturing process. Thus, every robot can

help this factory save 10 to 15 minutes time to finsh every fish cane manufacturing process. IN fact, time is money, because when every robot can help this factory to reduce 10 to 15 minutes time to compare human worker. Then, this factory can finish about 20 fish canes in one hour if it can use robot to help it to manufacture fish canes. Otherwise, if this factory still use human workers to help it to manufacture fish canes, then it can finsh about 3 to 4 fish canes in one hour. SO, the manufacturing efficiency ensures that robots must help this fish manufacturing factory to raise fish canes number more than human workers. So, in robotic behavioral economy view, manufacturing robots must help this fish canes manufacturing factory to raise fish canes manufacturing number and deliver increasing number to supermarkets to prepare to sell every day. Robots can help this fish canes manufacturing factory bring manufacturing time saving, rising manufacturing efficiency, improving performance and reducing wages expenditure long time advantages in micro economy view. However, manufacturing robots can also bring disadvanages to society, e.g. increasing unemployment ratio, increasing crime rate,

this factory workers will lose jobs and income, they need earn social welfare from government and increasing government finance pressure in short time, even long time in macro economic view.

Stanford University graduate program in economics, Scott lecturer explained that "in demand and supply economic theory for robots supply and demand case, robots supply number increasing may influence human workers demand number decrease. It sometimes calls " the efficient frontier".

No specific human beings were mentioned in any of economics classes. As robots supply and demand in market case, They (robots) may be purely theoretical " agents" who reached to the most reasonable sale prices in order to persuade any one businessman buyer to make manufacturing robot buying decision whether robots can help him / her to bring how much saving time , saving money, saving cost, improving performance, efficiency economic benefit before he/she plans to reduce workers number when he/ she decides to apply robots to replace human workers in his/her factory or office or any service department, e.g. cinema ticket sale service, shopping center customer service, shopping center cleaning , supermarket customer service etc. service or sale tasks. When robots can replace human to do any one of these tasks in any organizations. So, robots may be human worker agents who reached to prices the way robots would react to a software command. There was nothing that explained why some people thrived and others did n't or why truly brilliant, hardworking people could fail when much lazier folks succeeded." Having been admitted to the Stanford University graduate program in economics, Scott lecturer hoped to get his answers there.

How robots influence our future social changing? Using the right technology can be a boon to your business in this economy. For internet example, it is easier than ever to find well-matched customers all around the world, to stay in contact with them, and to more quickly design the products they want. If you focus solely on being cutting -edge, though you risk letting the technology
take over what should be very robust relationships with your customers , employees, and colleagues. IN nowaddays society, technoligical advances and cutomation, personal
relationships in business are more crucial than ever. I mean that robots can not replace human to serve clients to let them to feel more comfortable and passion more easily. For shoe shop case example, if the shoe shop apply one robot to serve its clients to replace human shoe salesperson to serve its shoe customers. Robots ensure that they can not persuade every shoe potential buyer to make shoe buying decision more easily when robots need to contact every shoe potential buyer. The reason is simple, because robots can not touch any one shoe buyer individual emotion very easier.
If the shoe buyer needs the robots to help him/her to choose any right shoe styles when he/she can not feel himself / herself can make the most

right shoe style choice decision. The robots can not replace human shoe salesperson to make shoe style choice judgement more easily. They must need longer time to analyze whether which shoe style may be the most suitable to the shoe buyer. Otherwise, human shoe salesperson may attempt to make the most right shoe style choice decision to help any one shoe buyer to chooce the most right style shoe because he/she owns shoe style sale experience, shoe style knowledge, the most important reason is that they can feel every shoe customer individual emotion to touch whether he/she will feel comfortable or happy when they attempt to help every shoe customer to seek the most right shoe style in every shoe customer whole shoe searching processing. Othwerwise, serving robots are only one machine, they can not touch or feel every shoe customer individual emotion whether he/she feel comfortable or unhappy or happy when they need to contact them in whole shoe searching processing. Hence, I believe that some tasks robots can

not repalce human staff to do very easily. Otherwise, robots may bring disadvanatges to let any one businessman to loss his/her customers, due to robots can not touch every customer

emotion to compare human staff in service tasks more easily. Robots serving customer behaviors may cause money lose and customers number lose to the shop in micro economic view.

Intellectual human economic behaviors

What does intellectual human economic behaviors mean ? I believe that when we choose or decide to do intellectual behaviors, then our societies will be influenced to bring economic growth in consequence.I shall attempt to indicate pollution case to explain how and why eithet our intellectual or foolish behaviors may bring economic growth or recession in consequence as below:

On one hand, for air pollution social case aspect example, if we only consider to buy cars to drive for working aimr or holiday leisure aim. Then, our societies air will be polluted. Our health will be influenced to bad. Our car driving behaviors may cause global environment air pollution serously. In long tiem, global air pollution will bring our bodies health to be bad. Although, ourselves car driving behaviors may bring our driving travelling leisure enjoyment and comfortable feeling in short time, also we so not need to pay public transport fare often, but we need to compensate ourselves health economic intangible loss due to air pollution , when cars number increases, dirty air will cause ouselves health to become bad.

In the result, we will need to pay more medical expenditure when we are old age, due to ourselves bodies will become bad, due to we breathe global dirty air every day, due to ourselves cars pollute air in long time, e.g. 10 to 20 years, even 30 more without limited air pollution environment. So, driving cars behavior may be one kind of human foolish behavior and our foolish behavior may bring ourselves future long time medical expenditure absolutely.

One the other hand, water pollution social aspect, if we often keep much rubblish to pollute sea, oil exploration porcessing pollute ocean , ships gas pollute ocaen, then fishes will eat polluted food and drive dirty water, due to global ocean is polluted.

In fact, because human only to conside how to buy boats to carry on leisure enjoyment activities, or catch cruises to travel on the sea. Also, oil manufacturers only consider researching anywhere to find new oil exploration places to manufacture oil product, when their oil exploration processes pollute ocarn . Consequently, global fishes drink polluted warer or eat polluted food. They will have poison. SO, human will have high chance to eat poison polluted fishes, due to fishes are poison or are polluted. So, human is doing foolish activities, we only hope to find oil exploration places to pollute ocean or we only spend money to buy ticket to catch ships to travel anywhere in global ocean. All of these human foolish behaviors will bring pollution to global ocean. On consequently, we will need to compensate to eat polluted or dirty or poision fishes, ourselves bodies health will be bad. In long time, we need have high chance to pay medical expenditure when we are old. So, pollution case may be one good example to explain how and why human foolish behavior may influence ourselves future need to compensate serious medical loss.

All of these human foolish behavior will bring pollution to global ocean. On consequently, we will need to compensate to eat polluted or dirty or poison fished , ourselves bodies health will be bad. In long time, we will have high chance to pay medical expenditure, when we are old. So, pollution case may be one good example to explain how and why human ourselves intellectual or foolish behaviors may influence future long time economic loss or economic growth or recession in micro and micro economic view.

On another water pollution aspect hand, if we often keep rubbish to sea, oil exploration processing pollutes ocean and ships' gas pollute ocean, then fishes will eat polluted food and drink dirty water, due to fishes will eat polluted food and drink dirty sea water because the global ocean is polluted seriously.

In fact, because human only consider how to buy boats to carry on any leisure water activities, or catches cruises to travel on the sea. Also, oil manufacturers only consider any where to find oil exploratin places to manufacture oil products from ocean, when their pol exploration processes can plooute ocean. Consequently, global fishes drink polluted water or eat direty food. They will have poison. So, human will have high chance to eat poison fishes.

Otherwise, such as pollutin case, it can infuence inflation or deflation. Consequently, the reason indicates supply and demand theory. If air pollution is serious, then we will consider health issue, global cars demand number may be influenced to reduce, when global cars number demand will reduce, global car prices and supply number will need to change to fall down in order to attract or persuade global car consumers choose to make car purchase decision.

Hence, global car manufacture number and car price will be influenced to reduce, due to global air pollution issue. Consequently, deflation will occur because when the country citizen usually does not spend much extra saving money to buy car expensive goods. Money value will be low. Otherwise, if global cair pollution is not serious, human considers to buy cars to enjoy driving leisure lives. So, global car demand is influenced to increase , also global car price will also influenced to increase.

Consequently, gobal human will choose to buy cars to drive. Due to we accept to spend extra saving to buy expensive car goods. Car sale price and supply may be influenced to rise up. Money value is influenced to reduce. Inflation may be influenced, due to global car consumers number increases, we would not have extra moncy to spend easily. Car expensive goods expenditure influences our spending habit to avoid to make car purchase decision more easily. So, human intellectual or foolish activities may bring inflation or deflation consequency in possible indirectly in macro economic view.

On conclusion, above pollution case explain that how and why human intellectual or foolish economic behaviors may bring inflation or deflation consequency as wll as economic growth or recession consequency as well as any goods demand and supply increasing or decreasing consequency. It implies that human behavior may have indirect relationship to influence any goods demand and supply number to either increase or decrease result as well as any goods price will be influenced to incrcasc or decrease in micro and macro economic view.

The relationship between social change and human behavior

Why does economic changes may influence human individual behavioral change? I shall attempt to indicate shopping behavior and staying at home behavior to explain their case and effect relationsip as below: Human behavior can be influenced by economic change or economic change can be influenced by human behavior? Why does recession may influence consumers reduce shopping desire? In social recession suitation, it is possible that many people lose jobs suddenly, due to businessmen lose many customers. They need to make decision to reduce employees number in order to continue to keep businesses. Consequently, many firms (organizations) their employees may lose jobs. When they have much time, due to lose jobs, they will feel to avoid to spend too much time and money to go to shopping often. Many losing jobs people, they will often stay at homes. So, they will reduce time to go to shopping, then non essential products won't their preferable choice purchase products. Hence, recession will change many losing jobs people their shopping or consumption desires to avoid to buy non essential products often . Usually when economic boom, many people have jobs to do because consumers number must increase when many people have jobs to do. Then, many people can accept to spend money to buy non essential products often. Many people feel spend time to go to shopping can satisfy their purchase of any kinds of new products useful psychology or desire. So, recession is one good example to explain it can influence many people do not like often to leave homes to go to shopping easily. Many people like to stay at homes, becaue they feel worry about spending too much shopping time when they leave homes. Their staying home time is one good negative shopping behavior example. So, economic change may influence human individual behavior changes , they have direct cause and efect relationship in behavioral economic view.

May human behavior influence economic change? Is it possible that human behavior may bring the country social economic change in macro economic or micro behavioral economic view ? I shall indicate publishing industry example. Do you feel that if there are many students feel learning is very important when they read many books or many of students feel interesting to read or they have reading new books in habit, then it is possible that the country will have many students like to spend time to go to any book shops to choose the books, they feel that they can help they learn new knowledge. Then the country will increase students number, they often spend time to visit any one book shop every week. Their visiting book shops behavior which may become their habits. So, the country will increase students number, they often spend time to visit book shops. Also, it implies that visiting book shops behaviors may be their behavioral habits.

So, when the country has many students often spend time to visit book shops , their visiting book shops behaviors may help any one book shop to raise books sale chance. So, the country's student individual often visiting book shop behaviors, their habitual visiting book shops behaviors must may assist help any one book shop to increase books sale number absolutely.

Consequently, any one book shop , its books sale bumber must be influenced to increase to increase because the country will have many students like or feel need visit book shops habit in order to choose any suitable books to buy to read at home in order to raise themselves learning effort. When the country has many bok shops often have many students visit their book shops, then their books sale number may be influenced to increase. It explain why student individual visiting book shop behavior may help any one book shop sale number increases also.

CHAPTER VIII

How human productive behavior may influence economic development

May any country which citizen behavior assist themselves country development? It is one cause and effect economic question. I mean that if the country itself citicen can not concentrate mind or energy to choose to do one kind of industry in order to let themselves country can bring the most benefit, then whether the counry itself economy can bring the most serious economic benefit. I shall attempt to indicate these countries themselves indistry choice to explain whether these countries themselves citizen productive behavior may help themselves countries to achieve the largest economic benefits. I shall indicate as below:

New Zealand farmer individual wine productive behavior

For New Zealand country example, this country concerns itself effort is foucs on farming agricultural aspect. So, this country has many farmers concentrate on farming agricultural aspect. May New Zealanders choose to spend time to produce different kinds of wines, e.g. wine or red grape wine is for the people are eating meat, or they are eating dinner.

When these New Zealanders their behaviors choose to do farming or agriculture to grow and produce different kinds of taste of white or red grape wine drinking products job. Themselves grape agriculture behavior will influence these New Zealanders themselves, they can learn how to improve different kinds of grape wine drinking products in order to achieve every kinds of white or read grape wines taste improving aim during their white or red grape producing process.

Why can New Zealander every individual white or read grape wine producers improve their white or read grape wine taste more easily? In behavioral economic view, it can explain that why any one New Zealander white or read grape wine producer can be encouraged or excited or persuaded to concentrate nervous and energy and effort to learn how to improve their white or red grape wine products easily.

In fact, New Zealand is one agricultural food export country. It has good natural environment resource , e.g. land, seed to provide any one farmer to produce themselves any kinds of agricultrual food products, e.g. fruit, or wine food products. Because New Zealanders know themselves country has enough natural resource . So, in common, many New Zealanders choose to attempt to do farming agricultural jobs in order to export themselves any kinds of fruit or meat or wine products to overseas or sell to domestic in order to earn profit.

So, when these New Zealand farmers number has been increasing every year. This country farmers will feel themsleves competition between this New Zealand farmers themselves are serious due to they may feel New Zealanders choose to do agriculture businesses in order to export themselves different kinds of farming food to overseas or sell to local to earn profit.

Hence, when many New Zealand farmers feel that farmers number has been increasing every year. They will feel themselves competition is serious. They must need to spend much time and nervous and effort to research what method is the best how to produce the best taste of white or red grape wine products in order to let local or overseas wine buyers to choose to buy his/her producing white or read grpae products to drink.

Hence, in competition psychological view, may influence many New Zealand white or reaad wine producers had been beginning to change their learning behavior on researching what method is the best in order to produce the best quality of taste red or white wine products to sell in order to attract overseas or local white or read grape wine drinkers to choose to buy his/her wine products. Their behavior will focus on learning how to raising or improving white or read grape wine taste method more than only focus on producing a large number white or red grape wine products. They believe wine quality is more important to compare wine producing number. So, New Zealand wine producers themselves wine producers behaviors have been changing on concentrating on researching wine quality method aspect more then wine producing number aspect in behavioral economic view.

America high technological productive behavior

For America example, US is one high technological country, it owns many high technological knowledge talent inventors, e.g. computer science inventors. Hence, US must attract many diferent countries owning high technological computer inventors choose to go to US to develop their computer science profession career. Also, it seems that when many computer science inventors or professions choose to go to US to develop themselves computer science new career. In behavioral cconomic view, due to their leaving themselves countries choice, which may bring influence themselve country job behaviors need to be changed. They must need to adapt US new live. Because they will forgive their past computer science job. These computer science professionals need to spend time to adapt US new lives. They " past computer science job behaviors" will need to be changed to their new US any computer employer's new computer science job model.

Because their traditional computer science jobs needed to be forgot in their themselves countries. They will feel their old computer science job knowledge and behavior needed to change in order to let their US any one new of computer company employer feels satisfactory to accept their new working behavior in any one US computer organization.

So, on the other hand, many US computer company employer will feel that they must need time to accept any one new overseas computer science professions their working behaviors, their working attitude daily, because these foreign comouter science professional, their past computer working behaviors and working attitude must be different to US domestic computer science professions.

In behavioral economic view, these overseas computer science professions, their working behaviors and attitude must be needed to change in order to adapt any one US new computer company itself domestic or local computer science professional stafs themselves daily working behaviors and attitude because these overseas and local computer science professionals must need to team work together.

In behavioral economic view, it is only one way that foreign computer science professionals must need to change themselves past country traditiona daily working behaviors and attitude in order to cooperate with these US local computer science professionals in teams more easily.

Consequently, if these foreign compute science professionals can change their past working behaviors and attitude to let any one US local computer science professional feels to cooperate with them easily in short time. Then, the US computer company itself whole computer professional teams themselves efficiencies will be influenced to raised or improved by the changing past working attitude and working behaviors of these foreign computer science professionals. So, in behavioral economic view, only if US any one computer company hopes itself computer teams themselves efficiency can be raised or improved when it decides to employ foreign computer science professionals and US domestic computer science professionals. They need to work in teams together. They must need to let these foreign computer science professionals to know how to change their working behaviors and attitude to let their domestic computer science professionals feel easy to work together. Then, the US computer company itself whole team efficiency must be rasied or improved easily in short time.

● China share market investing behavior

For China share market example, economic development depends on financial market. Because if many Chinese have interest to invest to carry on shares buying and selling activities in orde to learn how to earn shares interest and share profit when the China shareholder can make decision to sell himself/herself shares in the the high price, then he/she can earn money when he/she can sell the China company's shares in the high sale share price position.

If China has many Chinese like to spend time to carry on investing shares activities. Themselves shares buying and selling behaviors will influence China has many companies can increase fund from many Chinese shareholders in order to have enough money to expand or develop themselves businesses in China in long term.

Consequently, when China can have many Chinese like to attempt to carry on buying and selling shares investing behaviors in China share market. Themselves buying and selling shares behaviors can help many Chinese companies have effort to increase enough money or capital in order to continue to do their businesses in long term absolutely. So, it explains why when many Chinese become shareholders , they can assist China will have many companies continue to develop their businesses if many Chinese like to carry on shares buying and selling investing behaviors in long time in China financial investment market nowadays in behavioral economic view.